Energy Healing for Animals & Their Owners

An Earth Lodge® Guide to Pet Wellness

Other Earth Lodge Publications:

Natural Animal Healing:
An Earth Lodge Guide® to Pet Wellness

Grounding & Clearing:
Being Present in the New Age

Equine Herbs & Healing:
An Earth Lodge® Guide to Horse Wellness

To The Temples:
14 Meditations for Healing & Guidance

Energy Healing for Animals & Their Owners

An Earth Lodge® Guide to Pet Wellness

Written by Sandra Cointreau

Earth Lodge

An Earth Lodge® Publication
Gaylordsville, Connecticut

....

*** Energy healing is complementary to medical and veterinary care and enhances the healing process. Medical and veterinary diagnosis and care are always to be sought for significant injury or illness. ***

....

Your purchase of this book supports continuing research aimed to benefit the well-being of dogs, horses, and other animals. Thank you.

ISBN 978-0-578-03159-0

To my daughter, Maya, who is the shining light of grace, wisdom and warm humor in my life, and forever my greatest teacher.

Table of Contents

Preface

By Dr. Nigel Brown
Internationally Recognized Veterinarian

Mother Nature is truly amazing and mankind still only has scant knowledge of her power and mysteries. The energy of living is more than just what is in our food: the very essence of 'life' surrounds us all the time.

Diverse species such as swallows, salmon and whales are able to migrate unimaginable distances with pin-point accuracy; the air and sea carry immense swarms of insects, birds and fish that somehow communicate to perform synchronous movements and never collide; elephants hear sound through their feet.

Somehow animal life around us is able to hone in on a variety of different energy forms and achieve incredible feats that leave us mere humans truly astounded.

Theories of evolution place mankind and his ancestors on this planet, living in the same environments and evolving under the same influences as these other species. We share a high percentage of common genetic material with other mammals and share many physiological attributes. Our cells respond to the same external influences – Wolff's Law of bone structure changes (developed in the nineteenth century) states that bone in a healthy person or animal adapts to the loads affecting it. If stress on a particular bone increases, that bone will remodel itself over time to become stronger with the external cortical portion of the bone becoming thicker. The converse is true as well: if the loading on a bone decreases, it will become weaker. In a similar way, Davis' Law describes how soft tissue responds to demands.

Such mechanisms are mediated by the energy of local electrostatic fields with micro-currents that affect ion and colloid distribution in tissue fluid. Within the laboratory it has been shown that small electric currents applied to white blood cells cause cell regeneration but higher density current causes cellular degeneration – simple experiments which illustrate the influence of energy on living tissue.

Like my father, grandfather and great-grandfather, I was trained in the modern science of my day to become a veterinary surgeon in the United Kingdom. Newtonian principles taught that living organisms operate on a sequence of reactions between function, structure, bio-chemistry and electro-magnetic energy. There was no room in the curriculum for *airy-fairy* concepts of subtle energy and, sadly, this is still widely the case; today's state-of-the-art therapy for animals (and humans) consists of synthetic drugs.

Slowly, however, a broader range of therapeutic techniques is becoming acknowledged as valuable in our armory against disease and modern science is slowly developing the mechanisms to measure the subtle energy involved in these. Such equipment may provide some reassurance to those who require scientific proof before accepting what their own eyes tell them about reiki, homoeopathy, acupuncture, reflexology and many other complementary practices.

For the last 33 years I have worked to heal animals and help those who own and use animals to look after them better. Throughout my life I have been fortunate to see some remarkable sights involving the animal kingdom around the world; they have led me on an incredible journey which is still far from over. About three years after qualifying I came to realize that there was more to the biology of healing than the allopathic drugs promulgated as cutting-edge medicine and, with every passing year, that has become more clearly apparent.

I can still remember the dramatic success of treating my first patient with homoeopathy after a course at the Royal London Homoeopathic Hospital – my wife's life-long hay fever. The first pillule created a healing crisis within minutes - whole body itching. The next day's therapy produced the same result and she refused any more. Almost 30 years later she has never had another attack of her summer curse.

I still use homoeopathy and remain delighted with a therapy that is impossible to explain by conventional therapeutics where less than a single molecule of the original medicinal substance remains. However, work it does and the answer surely lies with an Einsteinian view of pure energetic healing which acknowledges that the electrons and atoms of the body are, in reality, bundles of energy. This, in turn, means that we are merely bundles of energy and so are the animals and objects around us.

From this standpoint it is not a big step to the realization that the energy of our bodies may not all be of the same wavelength and that some people may well have slightly different abilities to tune in to particular wavelengths.

So, if our bodies are composed completely of energy in one form or another, then the most fundamental mechanism to effect healing is to work on the energy that has been disturbed by the disease process.

Throughout my veterinary career I have always known that there was a power within me that allowed me closer connection with animals than many other people. I have seen and helped animals to recover from unbelievable problems - none more dramatic than the greyhound who impaled himself on a stake in Kent. He arrived at the clinic with chest and abdomen ripped open – heart, lungs, liver and intestines open to the atmosphere and ingesta oozing from an impaled gut. Three days later the same dog walked out of the hospital with a small section of rib and about four inches of bowel missing. His remarkable recovery was uneventful but while conventional wisdom said 'No chance', the subtle energy told a different story.

At a low point in my life I received a crystal healing session (somewhat skeptically, I admit). The results were truly awesome – both physical and mental – and, taking the divine guidance revealed at that time, I met my Reiki Master. Formally acknowledging Reiki as a potent healing force was a major step in that voyage of discovery, leading towards becoming a Reiki Master myself.

Meeting Sandra and reading her book has been another. She has a wealth of experience, gained over many years in several different areas of activity with subtle energies in both humans and animals, and channels energy into her patients using Reiki. In this book she shares some of her thoughts and experiences in this diverse and still controversial field.

I am pleased to say this is no repetitive manual of hand positions, nor an egotistical potpourri of successful cases. It is much more an exploration of energy, of healing animals and the way in which we can link with the energy within living beings to the benefit of mankind and the earth in general.

I hope you will enjoy reading her words, broadening your mind with her thoughts and endorsing the principles that the energy of our life force is an essential component of health.

Nigel Brown, BVSc MSc MRCVS MACVSc

Manama, Bahrain. November 2008

Chapter One

Vibrational Transmission

The Context of Energy Healing

In Japan, the life force energy is called Ki. In India, it is called Prana; in China, it is called Ch'I or Qi. The chakras are known throughout Asia as the locations where Ki enters and circulates. Branching out from the primary channel of energy flow at the chakras of the trunk, there are energy meridians related to each organ that reach outward along the limbs.

The energy-healing practitioner is someone who has studied, practiced, and attained the ability to contain and transmit heightened amounts of Ki. Energy healing increases the chakras' and meridians' capacities to accept and hold Ki. Energy healing balances the energy among the chakras and along the meridians, and it creates harmony within the body.

Energy healing complements medical and veterinary care. If an animal client has been sick or injured, the first step is to obtain a diagnosis and recommended medical and veterinary care. However, this so often is not enough and energy healing is used to support and expedite such care, or take over where that care can go no further.

There is so much that science does not understand. Atomic bombs arrange for the fission, or splitting, of the nucleic core of uranium or plutonium atoms to obtain huge nuclear explosions. One kilogram of nuclear bomb material can release the equivalent of energy of exploding 22 megatons of TNT. Quantum physics has proven that we are mostly energy and empty space... not matter. When your very sick client comes from medical and veterinary experts with negative information, use the new information carefully to focus your intent. But, when dealing with the mysteries of energy healing, the book *Course in Miracles* teaches:

> Nothing that you hear means anything,
>
> Nothing that you see means anything,
>
> Except the meaning that you give to it.

With time and experience, and repeated affirmations from healings, the energy-healing practitioner will learn that trust in energy healing is warranted and will learn to expect healings as a natural part of everyday life. With time, the energy-healing practitioner will learn to not attach fear-based meanings to what is said to describe the condition of the client.

There appear to be so many methods of energy healing in today's literature. Which one works best? Which is the true way? Perhaps, as has been said for many hundred's of years in Zen philosophy, "all the ten thousand things are one."

While there appear to be differences in techniques, in essence they are all the same. Lao Tzu, a famous Chinese philosopher of the fourth century B.C. wrote: "Thirty spokes, Share one hub. Adapt the nothing therein to the purpose in hand, and you will have the use of the cart." With energy healing, all techniques aside, adapt the nothing within to channel healing through your empty spaces.

In 1976, I studied transcendental meditation and for years thereafter I worked daily on my meditations. While it would be nice to say meditation initially was started for spiritual reasons, it wasn't. At the time, I worked at Arthur D. Little, a famous "think tank" in Cambridge, Massachusetts. My boss, in my first year-end review, told me that I was being productive, but not creative... and clients came to ADL for creativity. Quite chagrined at this criticism, and not at all convinced that it could be remedied, I read everything I could find about the creative process and learned that many inventors like Edison, Archimedes, and Newton said that discovery came to them in the moments of repose, following the work done to feed information into the brain.

Repose? That sounded too easy, but it was worth a try. I'd read that most laboratory scientists did repetitive experiments until finally they would suddenly see the truth emerge. And so, I practiced feeding data in and then taking time to meditate, even during working hours. At the end of the year, the company rewarded me with a major bonus of recognition. "Eureka!"

Think of the mind as being like a pail of water. When you stir it up, all the particles are in suspension. But, when the pail of water is allowed to stand still, all the particles float up to the surface. The same thing happens in meditation… intuitive and creative thoughts float to the surface during periods of quiescence.

In 1977, I was introduced to the art of energy healing by a Korean master of yoga, martial arts, acupuncture and energy healing. He'd been schooled for many years in a Japanese monastery. He was my "Sensai," an esteemed teacher in the ancient traditions. Through yoga, tai chi and meditation classes with him, I learned how to focus my energy and send it out the palm of my hand. Later teachers showed me how to redirect that energy and pull pain out of other people with the palm of my hand.

Shamanic journeying, which I started in 1992, taught me how to dream while awake and walk in the dream world (which Carlos Castaneda called the "parallel reality"). Shamanic journeying teaches how to use your intent to see into the lives of others, seek spiritual guidance, listen to your intuition, and ask for healings for others and yourself. According to anthropologist, Hank Wesselman, native shamans interviewed throughout the world agree that they travel to spiritual realms where they connect with inner sources of power and wisdom and link up with various spiritual helpers to conduct healings.

Anthropologists around the world have discovered that shamans from Siberia to Australia, from Amazonia to the northernmost Native American lands, follow similar methods of journeying to perform energy healing, including journeys to the realm of lost souls. (Michael Harner, anthropologist, author, teacher and founder of the Foundation for Shamanic Studies) While some native peoples use plants to help them reach a journey state, nothing but intent is needed. Most journeying done by westerners is readily enabled merely by the rhythmic, repetitive sound of an open-faced drum or gourd rattles.

The heart-beat drumming that so often accompanies shamanic journeying, like the rhythmic mantra chanting of meditation, reminds me of my earliest childhood lessons in transcendence, i.e., the act of simple repetitive praying of the rosary that had such power of calming and comfort.

As so many teachers have said over the years, including Carlos Castaneda and Deepak Chopra, a key factor in attaining the power to heal or moving in the dream time or experiencing full life actualization is "intent." Chopra, in his *How to Know God: The Soul's Journey into the Mystery of Mysteries* wrote, "The act of creation is reducible to one ingredient: intention...There are no magic tricks to making a thought come true, no secrets in miracle-working. You just intend a thing and it happens."

As author and energy healer Judith Motz wrote, "If communication did not exist with and among cells, tissues, fluids, and organs, then the vast bureaucracy of the body, with billions and billions of tiny parts, would never be able to collaborate to form a working organism." It is our spiritual non-physical self that directs much of our cell community's behavior, not merely our brain and its related nervous system. All of our cells are joined together in a vibrant flowing connected community. When we are in a state of relaxation and happiness, the cell community is able to operate in an optimal harmonious manner. When we are in a state of stress, with our thoughts in the tense state of spin, our cell community responds with tension and dis-ease.

People accept that radio waves sent from a transmitter many miles away can be received by them instantly, that cell phones and computer internet connections can receive instantaneous wireless messages, and that GPS systems can guide our cars from instantaneous transmittals sent from satellites in orbit around the earth. And yet, many of these same people doubt that their own cells can similarly receive instantaneous wireless communications from their own soul, or from another being. A wonderful book by Lynn McTaggert , called *The Field* discusses recent particle and energy physics discoveries that shed light on how messages are transmitted energetically in the body and also on research done to prove that people can perform remote sensing telepathically.

People know that they can turn their dial to 800 AM or 92.5 FM and obtain precisely the channel of news or music that they want, or they can dial their phone to 411 or 911 to obtain the operator or emergency assistance that they want. And yet, many of these same people doubt that they can turn their thoughts to a specific person, animal, plant or spiritual entity and instantly connect and communicate to that thought-dialed being. According to the teachings of Abraham, we can connect to anything that exists, because everything has been created through thought vibration by Source, and everything is a vibrational entity. (Hicks)

People have no doubt that GPS systems can operate across thousands of miles from satellites spinning around the earth, but most doubt that their thoughts also can transmit thousands of miles to exactly the human or animal channel they choose. Similarly, most people believe that their prayers to God or his/her high-level spiritual helpers (e.g., saints, prophets, angels) can bring them healing miracles. On the other hand, they tend to doubt that a healer can channel healing energy from that same God-Source to them over similar distances.

At different times over history, people have felt strong connections to various high-level spiritual entities, all of whom are part of God-energy. Some pray only to the Great Creator (known as God, Allah, Wanka Tanka, Source and by other names). Others pray to high level intermediaries with whom they feel a special affinity, or whom they feel may have more time to be of support to them (such as the Blessed Mother Mary, Buddha, Jesus the Christ, Shiva, Ganesh, White Buffalo Woman, and a wide range of saints, yogis, and other holy guides). Some have difficulty believing that specific spiritual entities can simultaneously answer prayers and provide help to so many. And yet, these same people don't doubt that every person on the planet can connect at the same time to Google on the internet and get an instant response. In this new time with our new scientific understanding, it is time to be more comfortable with our essence as vibrational beings. We have such magnificent complexity in our God-made bodies and their vibrational energy fields, which are many times more powerful than any human-made cell phone, radio, laptop, or GPS!

People accept that certain types of atoms can be split to create massive atomic explosions, and yet they doubt the energy within their own body's atomic structures. Particle physicists have proven that our atomic structure is mostly made up of energy. If all of our cell's subatomic particles were put together, they would fit on the head of a straight pin. We are vibrational energy-based beings.

Our vibrational nature is well described by Abraham, a group of Source entities channeled since 1988 through Ester Hicks. We see vibrational light waves as colors, we hear vibrational sound waves as noises, we sense temperature and humidity vibrationally through our skin, we experience love vibrationally in our entire body when caressed. In the same way, we receive informational vibrationally through our telepathy. We all experience the receipt of phone calls after we think about someone, and yet many of us doubt the existence of telepathic messaging. (Hicks)

As humans, we are often doubters. To counteract our doubting mind, it is useful to create validation opportunities that are personally convincing. We can practice sending thoughts to a friend, record our dreams and see their predictive validation, ask a pet owner if a message being received seems relevant to their life, share our telepathic information and see if it resonates with the person it is about. As we share, we will obtain validation after validation. In time, confidence develops and doubt disappears. With faith in this ability, our potential becomes limitless. Intuitive, telepathic, and psychic ability is available for everyone to enjoy. It is a skill, not a gift. Like all skills, whether playing a musical instrument, painting a picture, riding a horse, or skiing down a mountain slope, some will be better at it than others… but everyone can learn. Everyone can develop intuitive, telepathic, and psychic ability through focused effort.

Some people are more open to developing these skills. They may have had family members with these skills, and childhood witnessing of validated telepathic messages made them receptive. There is some evidence suggesting that people who were abused as children or brought up in dangerous living conditions may more readily develop intuitive skills. They may have had to subconsciously keep their intuitive vibrational antennae active as a matter of survival, on heightened alert, to anticipate subtle differences in behavior and forewarn of possible abuse or danger.

For decades, since being a teenager, I had experienced telepathic communication with people and prophetic dreams about events. I'd come from a long lineage of women who accepted telepathy and prophesy as a normal part of everyday life. I'd witnessed events in dreams hours before they happened, and witnessed events in real-time moments before they happened. I received prophetic dreams about my animals' health, which I viewed as messages from the high-level spirit realm. But, I never imagined that animals communicated telepathically. In 2001, a friend that was studying animal communication, helped us to locate a lost pet. Despite this amazing experience, I still was skeptical about animal communication. I decided to read Martha William's useful book, *Learning their Language*, and practice the exercises in it. As part of the practice exercises in the book, I started to put questions to the animals I met. And, much to my surprise, animals started to answer my questions. Only sharing the answers with the animals' owners eventually convinced me that these answers were valid. Now, no matter how bizarre the answer, even about their past lives, I have come to accept all answers received from animals as likely to be true. Like a prophetic dream, versus a normal dream of rehashing the day or working out an anxiety, a true message feels quite different and one learns to discern the difference. Sometimes, I have asked for physical confirmation of an answer, and repeated the question several times with the animal shaking its head slowly "yes" each time or giving another signal each time to confirm an answer.

To fully embrace the amazing abilities of your full physical and non-physical being, simply live your life by practicing the art of the possible. Be open to all that you can be. Be faith-based, rather than fear-driven. Being a great energy healer involves a total immersion in positive thinking and manifestation, with pure intent for the healing to manifest.... in the ancient healing tradition of: "so shall it be, with harm to none and good for all."

Chapter Two
Quieting the Mind

The Foundation of Energy Healing

White Eagle wrote: "Wisdom comes to those who are calm and tranquil in spirit, to those who wait upon the Lord... Your spirit is part of God, and all knowledge lies within you. If, in your meditation, you will go deep within, you will find the center of truth and of the infinite powers which await man's use. You will touch the spring of all happiness and health."

The foundation for obtaining pure intent is the quieting of the mind. There are many ways that work for quieting the mind. It is valuable to know these many ways, even though they all help you obtain the same end point. Sometimes, when our minds are particularly chaotic and filled with noise, we may have to try several of them until one that suits that day provides the necessary quiet.

This book and the associated training courses cannot replace a lifetime of reading, experience, or effort. We must all do our own work to learn how to maximize the art of quieting our mind so that we can be optimally joyous in the moment, manifesting our visions, including our intentions in energy healing. Whether we start with the study of Buddhist meditation and chanting, contemplative repetition of the Catholic rosary, or vision questing as part of the study of Shamanism, we will learn to find those spaces of time where we are fully in the now. Each book and class that we experience adds to that learning. Because we are each different, the way that we approach and conduct deliberate intention and related energy healing will be different. Nevertheless, each of us has the ability to provide the comfort and release from disease that energy-healing enables... and even to save lives. Below, are some well-known meditation techniques.

Repetition of a word that is soothing eventually leads to a cycle of thought and no thought, thought and no thought. In transcendental meditation, we are given our own personal mantra to repeat to bring on the state of quiet. Some Buddhists use the word "Ohm," which is almost a mid-range hum. Tibetan Buddhist monks make a deep guttural sound that seems to cleanse the atmosphere. Gregorian chants are more melodic and varied. Simple concentration on breath, in and out, slowly, rhythmically, works for many meditation practitioners.

I like to use the simple repetition in my mind of the words "No Thought," three times as I breathe a long in-breath and three times with a long out-breath. I also like to visualize breathing in clean, fresh oxygen to every cell, bringing that fresh, rejuvenating breath into the most remote cells in my toes and fingers, as if my arms and legs are hollow tubes. On the out-breath, I like to visualize that particles of toxic elements and negative thoughts are being pushed out from each cell, as I give particular attention on long, powerful out-breaths.

Eckart Toile's many books and CD's on the "power of now" are excellent in helping you to be present in the moment. After much practice, you will be able to notice when you are in spin mode and be able to hit the pause button on your spinning thoughts. When our mind is spinning with thoughts of guilt over something in the past, shame over something that once occurred, anger over some grievance, anxiety over something we fear in the future, simply hit pause and spend a few conscious moments with your breath coming back into the present and become focused on something that you can constructively and positively do to fully engage your mind in the present.

Meditation is being in the present. The optimum meditation state of "no thought" will give you and each of your cells the most restorative and healing rest. Each cell will be able to go into a state of relaxation and harmony. Your mind will be like a quiet pool and the creative sub-conscious thoughts and brilliant spiritual guidance available will be able to rise to the surface of the pool, for you to receive.

With practice in meditation, we can reach a state where our bodies seem to disintegrate in peaceful bliss and there is no noise in our head for some extended minutes. This is not a trance state. Just the opposite. It is a state of heightened awareness... intense presence in the now. We experience heightened awareness of every sound, movement, and temperature change, around us during this state of "no-thought." We note the brief thought of awareness of a change, dismiss it and return to the condition of "no thought." The experience brings on total relaxation and personal healing. As our practice continues, we may feel energy moving in a circular motion through our body, round and round, cleansing and healing with each cycle.

For many, guided meditations are the easiest way of developing their ability to meditate. CD's are available with guided meditations. Group guided meditations, in person, have a special synergistic ability to enhance a guided meditation. The guide's voice is able to hold the attention of the group and keep them in the moment, as they listen and wander along the imagery of the guide's meditation as it is being conducted. In guided meditations, participants have some thought, as one travels on the journey that is being guided… but participants are very much in the now. A guided meditation might be an evening walk along an ocean beach, with the sound of ocean waves and sea creatures played in the background by the meditation guide. It might be a morning walk along a stream, with the sound of morning song birds played in the background. It might be a celestial climb through various levels to temples in the sky, with angelic singing or Tibetan bowls playing; or a downward climb through caves to reach a deep, inner, crystal cave accompanied by heartbeat open-faced drumming. Every now and then the guide will note that you are walking in a certain place, perhaps arriving at a meadow or special outlook, perhaps meeting a special animal or finding a special crystal. The purpose is to let your mind be in the meditation and wandering along the path, letting your own mind provide you with the experience that is relevant to you. The key is that you are in the now – and out of the spin mode of racing thought. Some beautiful and healing meditations are available in Maya Cointreau's book *To the Temples:14 Meditations for Healing & Guidance.*

Yoga can enable practitioners to achieve a meditative effect of "no thought" by the concentration required to hold a series of postures together with control of breath. The postures are balanced and each sequential posture leads to a harmony of movement and breath. As with transcendental meditation, the experience brings relaxation and a sensation of the energy flow within the body, moving in a circular pattern, that provides healing with each movement.

Tái Chi Chúan also enables achievement of a meditative state. Tái Chi Chúan or "Supreme Ultimate Boxing" is based on the principles of the I Ching and the philosophy of Lao Tzu, developed over 2400 years ago. Tái Chi is similar to Yoga in that it involves a harmonic fluid rhythm of balanced movements and breath control in a steady yin and yang, to and fro, equilibrium.

Inherent in the sequential 150 sets of movements, Tái Chi adds an important element toward learning the art of energy healing for, with many of the sets of movements, there is a thrust of the arm and hand. Often the hand is moved forward with the palm facing the shadow opponent (for most often this subtle form of martial arts is practised quite alone and quite slowly). With the thrust of the palm, we learn to send out our energy like a laser beam. In opposition to the thrust, there are returning movements where the hand gathers up energy, pulling it back to oneself.

Similarly, Qigong is a Chinese branch of learning that is said to be more than 2000 years old. It deals with the movement of the vital energy flow of the body, including ancient methods of inhaling and exhaling, to promote mental calm, clarity, and well-being. Qigong involves physical movements, but focuses primarily on control of breathing while in certain postures as a means of building up Ki. As with Tái Chi, Qigong prepares the healer by quieting the mind, focusing intent and increasing awareness of the body's energy flow.

Carlos Castaneda describes a series of ancient shamanic postures from Mexico, learned from his teacher, that are similar in concept to Tái Chi. They involve flowing postures conducted slowly and in a balanced manner, like Tái Chi. They focus more, however, on pushing out and gathering in of energy. Castaneda describes these postures in his book *Magical Passes*.

While all of these methods are directed at the concept of quieting the mind, focusing intent and learning to direct energy flow, other activities can also lead many people to being in the now. Many pastimes such as jogging, rowing, skiing, carving, cooking, gardening and needlework, if done in a rhythmic, balanced, purposeful way, can lead to an awareness of the circle of energy flow in our bodies, to our internal rhythm, directing our energy outward and pulling the universal energy inward. Some people enjoy extreme sports, not realizing that some of the wonderful feeling they experience derives from their concentration requiring them to be fully in the now.

A Simple Breathing Technique to Build Ki for Energy Healing (from Diane Stein)

Follow the sequence below for a simple and relaxing meditation that will clear, strengthen and harmonize your energy flow.

- Sit quietly in a chair.

- Focus your attention on your navel area, called the Hara.

- When the area begins to feel warm, move this warm energy downward to the perineum, called the Hui Yin. (The Hui Yin point is between the genitals and anal area.)

- Move this warm energy upward along the back to the kidney area, called the Ming-Men. Hold it for a few breaths.

- Raise this warm energy further up the back to the crown. Hold it there for 5-10 minutes.

- Bring this warm energy downward to the brow area, often called the Third Eye, where the pituitary gland is located.

- Flow the warm energy steadily and slowly downward to the navel area, the Hara. Hold it there until the warm builds up again then repeat for several cycles.

- After practicing the above and developing the ability to increase the number of cycles to more than 10, start to include the legs.

- Send the warm energy from the Hui Yin downward along the back of the thighs and calves to the underside curve of the foot, called the Bubbling Well. Hold it there for several minutes.

- Flow the warm energy, steadily and slowly, upward along the insides of the calves and thighs to the Hui Yin.

- Complete the exercise by grounding the energy at the Hara, rubbing the navel area in a spiraling motion (for most women, counterclockwise then clockwise) (for most men, clockwise then counterclockwise)

Chapter Three
Deliberate Intention
The Method of Energy Healing

Changing Perception

There are many valuable books that help us to change our perception of people, situations and opportunities. For examples, look to the writings of Lao Tzu, the Dalai Lama, Bear Heart, Fools Crow, White Eagle and Abraham. Most spiritual teachers, ancient and modern, have similar messages. They teach compassion, forgiveness, humility, generosity, detachment, trust in miracles and respect for God's mysteries.

From 1977 to 1979, whenever I had any problems, my Korean Buddhist Sensai would tell me to say "Thank you, God, for this wonderful opportunity to learn." In 2005, many years later, I read the autobiography of Bear Heart, a great Native American medicine man. He said "The way my people take care of something that we're not happy with is to honor it and say [to it], 'Thank you, you've taught me a lesson'." He also said "Be grateful for all the difficult situations in life because you can learn something from each one."

White Eagle wrote "nothing happens out of order or by chance, and the great Law brings those very conditions in your life which you need for growth. So accept all that happens. Look for the lesson that has to be learnt from the experience."

I spent 4 years meditating daily on the 365 step-by-step meditations in the book *Course in Miracles*, during the mid-1980's. I repeated the Course again in the mid-1990's. The Course is excellent for shifting one's thoughts toward positive thinking and realization of miracles. While doing the Course, miracles become a daily part of life, as one changes thoughts from "fear" thoughts to "faith" thoughts. The *Course in Miracles* teaches you to look on your worst enemy or the person that has hurt you the most as "your Savior," because that is the person that has brought you your greatest opportunity to learn. This teaching allows you to be able to forgive your enemy and the Course teaches that "forgiveness is the key to happiness." What is not stated, but true, is that forgiveness allows you to get out of the spin cycle of negative thought and enter the 'in-the-now' cycle of positive manifestation.

Abraham, the collection of source entities being brilliantly channeled through Ester Hicks for more than 20 years, teaches that encountering adversity gives us strong feelings of what we DON'T want and sets up strong feelings that are referred to as "rockets of desire." Over time, these rockets of desire accumulate in our special savings account of desires, which Abraham calls our "vibrational escrow." For many, these rockets of desire stay in vibrational escrow, rather than manifesting, because our minds are spinning about negative things that have happened, our feelings of anger or distress are keeping us from focusing on our desires, and our expectations are not aligned with our desires. (Hicks) But, once we line up our thoughts toward positive expectations and put ourselves into a happy, relaxed state of non-resistant harmony, that vibrational escrow of desires can quickly manifest. The key is to focus on conceptualizing the framework and details of our positive desires, rather than holding our thoughts in spin cycle about the negative current or past situation. Hit that internal "Pause" button to stop the spin, and do something light-hearted to shift moods into a better feeling state. Think "wouldn't it be nice if…." Focus on part of the cup that is half full, naming each of the positive aspects in that half full cup.

Forgiveness is part of the process of putting aside negative feelings and moving into a space where positive deliberate intention can successfully manifest. Even though you might not love that harmful thing someone has done to you, forgiving with love in the heart is simply recognition of the person's creation by the Higher Being for some higher purpose…as we all are here for our higher purpose. The Dalai Lama wrote: "Happiness comes from kindness. Happiness cannot come from hatred or anger." (Dalai Lama)

Deliberate Intention

To manifest a deliberate intention of healing, one must first develop a vision of a desired outcome from the energy healing. Obtain information from the client on how they or their pet are feeling and discuss how the dis-ease developed over time. Try to explore any associations that the patient has with the part of the body that is not well, or about the type of illness being experienced. Discuss the patient's expectations in terms of well-being and try to understand whether there are negative learned or familial expectations that might create resistance to healing within the patient. Communicate about the patient's vision for well-being.

It helps to tap into our intuition before beginning an energy healing. This could involve telepathically communicating with the patient in advance of the healing session, to ask about their illness and their desires and expectations. It could involve asking for a dream that provides information, conducting a shamanic journey to obtain information or meditating for guidance.

Write in your journal about the information that is received through telepathy, dreams, journeys and discussions. If guidance is received on how to conduct the healing, you may share this with the client as part of obtaining permission. Then the path of healing can begin as a fully shared effort, without resistance. With clear and mutually developed intention, whatever healing is conducted is then most likely to be embraced and supported by the patient.

Each energy healing technique involves some ritual, because ritual helps to put bring us into the now and hold us in the present. Feel free to add to the basic rituals that are taught about your healing methods. Add sound, aroma and visuals that support maintaining your intent and your patient's calm and harmonious participation.

Our breath is a powerful tool for focusing our intent. A Zen Master artist will often contemplate a brush stroke while inhaling, and execute the brush stroke with great intention while exhaling. The martial arts master may contemplate an attack move while inhaling and moving into position and then execute the attack move while exhaling. During energy healing, it is powerfully effective to use your exhaled breath to support your transmission of energy to the patient, use your exhaled breath to draw your healing symbols on your hands or use your exhaled breath to release extracted dis-ease. While many people conduct energy healing as passive conduits, grounded application of your personal power of deliberate intention can intensify the healing.

The Ethics of Energy Healing

No one here on this planet is perfect. We are all here to learn, be tested, create and grow to attain our higher purpose. And so you are not expected to be perfect in order to do energy healing. The main ingredient for healing is that you have pure positive intent. As part of the clarity of your intent, it is appropriate to also follow some well-defined ethical guidelines when serving as a channel of energy healing for others. The following ethical guidelines are generally appreciated:

- Recognize that your healing energy is not derived from our physical power, it belongs to the Great Creator and is passing through you. Treat it with the respect that it deserves as one of life's great mysteries.

- Recognize that optimal healing occurs when the patient and the healer work together toward the desired state of well-being.

- Always prepare yourself and your patient with a calm confident approach to the healing process, a request for permission to proceed and a prayer of protection.

- Empower the patient to heal themselves and assist them in their personal growth and in learning self-healing techniques.

- Do not use mind-altering alcohol or drugs before or during energy healing.

- Never do energy healing when feeling negative emotions such as anger or depression.

- Never make sexual comments or touch the patient in a sexual way. Energy healing is done on clothed patients.

- Monitor your views of a patient's condition and needs to be sure that you are not projecting, rather than intuitively reading, their true condition and needs.

- Never violate free will by imposing energy healing on those who don't want it.

- Do not take on the patient's disease, your own body is sacred and not a trash bin for misplaced problems removed from others.

- Acknowledge that helping people as a healing practitioner can be tiring and warrants rest and replenishment between healings.

- Thank the Great Creator and all of your helpers for the blessing of each healing experience, and thank Mother Earth and Father Sky for any disease you toss them to remove.

- Respect and value all energy healers, regardless of lineage or technique - all are doing their best to improve the lives of others.

- Encourage clients and students to use their intuition in selecting healers and teachers and always empower them along the path of their choice.

- Be truthful in explaining your healing training, practice, techniques and experiences.

- Respect and value traditional medicine and other forms of healing as complementary and consistent with energy healing. Any negative thoughts about physical medicine are points of resistance.

- Never contradict the recommendations of licensed health care providers or suggest a client change prescribed medicine or treatment. To fully believe in the miraculous potential of energy healing is to embrace healing with and, if necessary, around allopathic healing activities.

- Complete the healing process with a grounding and protection prayer and/or ritual, so that the patient is fully present in the now and connected to earth energy.

Students of the healing arts often want to know whether they should attempt healing when they are not feeling well. The answer to that depends on the issue affecting their well-being. If the cause of not feeling well is emotional, such as anger, depression or mental illness, then healing should not be conducted. However, if the cause of not feeling well is purely physical, healing may be conducted through the channeling technique.

As a sacred channel for energy healing, such as by Reiki or Healing Touch, it is not necessary that your own body feel completely well before a healing. Energy healing techniques assume that healing energy comes from the Great Creator, whether called God, Allah or Source, and that it passes through the healer. As long as your intent is pure, the healing energy will pass through you to the client with positive effect for both you and the client.

Students in the healing arts wonder whether they may call on supports from other than Source or the Great Creator. Some wish to call on saints, angels, yogis, totems or spirit guides as intermediaries to facilitate bringing in the healing energy. All of these spiritual entities in the higher level spirit realm are connected to God-energy. To be comfortable with your choices, simply call on your higher power, your own internal link to Source, for confirmation that the spiritual entities you are asking for support are right for you. If your response is positive emotionally, and you feel good in your gut as a devoted healer, include them in your community of healing helpers.

Others conduct shamanic journeys to non-ordinary, parallel reality where they call on their Totems and Guides to do the healing work. If the healing effort or journey turns an unexpected corner and you find yourself having to use your personal power, ask your Totems and Guides for help in your own cleansing and healing before you leave the non-ordinary, parallel reality. Typically, the channeling of healing energy or participation in a healing journey will result in a cleansing and healing of your own health problem.

It is important to concretely seek the patient's permission to do a healing. With animals, it is often best to stand with your hands open to see if the animal comes to lean into them. When the animal moves away, it is usually a signal from them to pause or stop the treatment. With animals, submission and relaxation into the healing energy can be perceived as permission. Also, trust your telepathic abilities of medical/veterinary intuition.

With people, you need to specifically ask: "Do I have your permission to do energy healing?" or "Do I have your permission to do a journey (vision quest) for your healing?" Sometimes there is an intermediary asking for you to provide healing for someone they love or are taking care of. In such a case, it is appropriate to say that you will, so long as the intermediary requests and obtains permission from the person.

Are there conditions when a requested and desired healing should not be done? I would say no. Only fear and lack of a clear positive intention on the part of the healer can interfere with a positive healing. Sometimes healing practitioners are fearful of doing healing when someone is bleeding or during surgery. They are concerned that affecting the energy flow could lead to more bleeding. Any healer who feels fear and is unsure should not proceed with a healing. I believe that healing can be done at any time, in any circumstance... the only thing that matters is your faith, confidence and intent. With clear, focused intent, nothing bad could ever come from a healing effort.

Healing Touch, practiced by many nurses and doctors, has developed an International Code of Ethics for Healing Touch Practitioners. It is relevant for all energy healers to review and apply this ethical framework. See: http://www.healingtouchcanada.net/ethics.html

The International Center for Reiki Training publishes a Karuna Reiki® code of ethics. See: http://www.reiki.org/ and all registered Karuna Reiki® Masters are obliged to follow it.

An Affirmation

Early in 2005, I had a very bad fall on a concrete sidewalk, breaking my nose, knocking out a tooth, getting two black eyes and seriously hitting my head. Blood was everywhere and I felt that I was soon to die from the injury to my head. The sidewalk filled with people, all afraid to touch me because of the head injury and all the blood. They were screaming into their cell phones and two ambulances arrived. In the meantime, lying on the sidewalk, initially I was afraid to do energy healing work… afraid the bleeding might be increased. Within seconds I decided not to be fearful and asked all of my cells to stop the bleeding. While on the sidewalk, I recited the Reiki symbols over and over in my mind and used my breath to slow down my heart rate and blood pressure.

CAT scans at the hospital showed that the head injury had caused a brain hemorrhage, but that it had quickly stopped and coagulated. That day, my years of practicing various forms of meditation to control my thoughts from going into a negative spin cycle were rewarded. I feel confident that the immediate decision to do energy healing and to acknowledge the role of each cell in my body, despite the massive flow of blood all over the sidewalk, saved my life.

Chapter Four

Gathering & Building Your Energy to Heal

Each method of energy healing relies on:

- Visualizing that the healing will be realized.

- Trusting in the healing process.

- Maintaining focus on manifesting the healing vision.

- Honoring your healing guides and teachings.

Bear Heart, a powerful Native American Shaman, suggested this ritual: "Lie down with your navel toward the Earth and your head to the north, saying: 'Grandmother Earth, please send your healing energy through this body and bring it back into balance."

Most healers engage in ritual to gather their energy. Ritual prayer, whether through bowing to Allah at certain times per day and reciting the Muslim prayers, through walking the Stations of the Cross in a Catholic church or through daily chanting of Buddhist prayers, helps us to center ourselves and gather our energy.

You should feel comfortable with creating your own ritual prayer system, and regularly adding to it or changing it. All prayer is good. We all pray to the one Great Creator, the Source of all that is, only the names and procedures are different. As a Catholic, I have always enjoyed the many rituals available in Catholicism and appreciated the serene beauty of churches for these rituals. But, for many years, I also enjoyed a daily jog of several miles, where I recited the rosary as I ran along.

It is wonderful to create a ritual that can be done in nature honoring the Nature Devas and calling on all Spiritual helpers that work with me. My friend and teacher, E. Barrie Kavasch writes of her morning ritual of walking her medicine wheel garden and laying stones at the prayer flags, describing her ritual and the creation of the garden in her beautiful spiritual and informative book, called *The Medicine Wheel Garden*. Some healers create a circular labyrinth in their yard and do a daily walking meditation to the central altar and back out again. It doesn't need to be done with hedges, but a simple labyrinth pattern in stone or brick can be created.

Many Native American traditions involve honoring the spirits and totems of the four directions, as well as the four colors of earth peoples. For my own personal ritual of doing this, I've created altars at north, south, east and west places circling my home. I walk the circle and call on the ancestral guides of the white, red, yellow and black peoples, and call on the elements of air, fire, earth and water, each in turn. I light white, red, yellow and black candles, each in turn, one per altar. Having the altars around the house, near the property boundaries, creates a sacred protected space. Sometimes, I stand in the center and smoke a redstone prayer pipe filled with Native American healing herbs, such as mullein, sage, yarrow, bearberry, sage and some tobacco. The prayer pipe is smoked in the four directions, honoring Father Sky, Mother Earth, and Father/Mother/God by sending the prayers on the wings of the smoke to the heavens. Many times after this ritual, I take photographs, as this is often a time when beautiful orbs of light beings will appear in the sky, many times hovering over each of the altars and their burning candles. (Some wonderful photos of orbs and energy can be seen on http://www.edenisnow.com, in the archives of questions and answers) For the next several hours, it is lovely to look out of the house and see that the four direction prayer candles are burning and sending up my prayers to the heavens.

Visualizing that the Healing will be Realized

Whether riding a bike or driving a car, it goes in the direction that your eyes turn. When playing tennis or golf, the ball goes where your eyes look. The energy of the body aligns with our intent. In all things in life, we eventually manifest our conscious vision… and our sub-conscious expectations. If we spend a short time each day in our prayers and articulate our desired vision but then spend the rest of the day worrying (which is a form of doubt) and having negative expectations, we are not likely to manifest our vision. Energy healing involves highly focused manifestation of a healing vision, and requires the healer to be positively in the moment with high expectations.

Personally, I use many methods to focus my vision on a client's healing. Some of these are:

- Choose crystals and colored candles that seem to address the needs of the client.

- Draw a picture, pick out a symbolic tarot card or use a photo of the client in a healthy happy state.

- Put the name of the client in the altar, with their own written affirmation, to continue support to a healing after it has been conducted.

For example, I wrote an affirmation to place on the altar that I created for Rupert, the 7 month old puppy, who had failing liver, failing kidney, leaking veins and edema, as part of a battle with infection. The affirmation said, "I ask all of the universal healing powers for Rupert to fully heal. This is my faith and will. So shall it be." For Rupert, I placed the affirmation on top of an ancient healing symbol used in Reiki, called the Antakarana Symbol, along with crystals and candles. Rupert is now a healthy, happy service dog who works daily with special needs children.

An ancient symbol called the Antakarana Symbol was used in Tibet for meditation and healing, reportedly for thousands of years. (Stein, Rand). It is useful to place the Antakarana Symbol on your altar, particularly for distance healing. The International Center for Reiki Training sells an attractive laminated version. I've done a little watercolor of my own version of the symbol, which you also may do.

Native Americans refer to the vision quest as a tool for insights and to direct the course of life. The shamanic healing practitioner journeys to a place in non-ordinary reality for a vision. During shamanic journeys, healing guides help to provide insight on the nature of a client's disease, they may even note specific herbs to be given and how they should be prepared. Sometimes the healing guides along the journey show methods to extract an illness and describe crystals to use to heal the disease. (Harner) During a visionary journey, the shamanic healing practitioner may recover lost spiritual pieces, sometimes referred to as lost soul pieces. (Ingerman) Soul loss can occur during traumatic events, abused people may let pieces of their soul go into safe hiding for personal protection, people that are critical and jealous may take pieces of soul. People intuitively acknowledge a sense of soul loss when they comment that, during a certain relationship, they felt that they kept losing pieces of themselves. Once gone, few know how to recover these soul pieces in order to be fully whole and strong and this is work that only more advanced shamanic practitioners are comfortable to undertake.

Trusting in the Healing Process

Universal truths are available for us to know, and universal knowledge is available for us to receive. We have only to tap into the flow of universal knowledge through our intuitive spiritual non-physical self and know all. (Wright) We may know something non-physically and be able to reflect it through the actions of our physical self. But knowing may still not enable us to explain all that we know. Michael Adam wrote in his book, *Wandering in Eden*, "There can be no serious answers to questions about the meaning of life, for to ask about life is to stand back from life and pretend one is not it."

Just as we might not be able to articulate the mystery of life's meaning, it is difficult to explain the mystery of what makes energy healing work. It is our "knowing" through personal experience that affirms to us its existence. We observe that energy healing does work and that others have been witness to our healing events. We see its reflection in the physical. We know how to prepare and conduct the healing but we operate on faith in the mystery of healing, rather than analytical explanations.

The painter does not paint the light but only the things upon which the light alights. The geese that fly over the water do not put their reflection on the water and the water does not create that reflection; yet the reflection is there and the painter can paint the scene. (Adam) The energy healer does not put his or her energy into the client but, as a hollow conduit, the healer enables the healing energy, mysteriously and inexplicably, to flow from Source into the client.

In his teachings of Taoism, Lao Tzu, more than 2,400 years ago, wrote, "The Tao that can be told is not the eternal Tao. The name that can be named is not the eternal name. The nameless is the beginning of heaven and earth. The named is the mother of the ten thousand things." (Feng)

Lao Tzu also wrote about the unseen, the non-apparent:

"Thirty spokes share the wheel's hub;

It is the center hole that makes it useful.

Shape clay into a vessel;

It is the space within that makes it useful.

Cut doors and windows for a room;

It is the holes which make it useful.

Therefore profit comes from what is there;

Usefulness from what is not there." (Feng)

Hank Wesselman wrote "It's not about clearing up these mysteries. It's about making the mysteries clear. When we experience the mysteries directly, we make them our own."

Bear Heart wrote: "We were all given the same amount of spirit. None more, none less. The difference between individuals is allowing the Spirit to have more of you… yielding to that spirit more."

To grow in your understanding of energy healing, study books about healing, such as Fools Crow's books, and take courses, making efforts of telepathy and energy healing and accepting the affirmations that they bring. After some training, it is wonderful to do healing-shares, where a circle of energy healers come together to work on each other and get feedback from healing colleagues. Create and participate in shamanic journey circles, where the monthly effort of journeying and sharing enriches everyone's abilities. Dream circles are also very effective, where each participant brings one dream to share and try to interpret. Witnesses to the journeys and dreams, and the journals kept, create a record of affirmations for those prophetic messages that come true. More experiences bring more affirmations, until your confidence in Spirit's willingness to cause "miracles" every day at any moment grows to complete faith.

Working with animals enables the affirmations to be quite clear. Animals are free of intellectual bias and usually are open to healing. When an animal heals, against all odds, in defiance of all professional opinion, it is an affirmation. For each animal healed, it is a mystery. But, for the healing practitioner and for those witnessing the repeated events, the affirmations pile up and trust becomes resolute.

As an engineer, a scientist, I believe that nothing is really needed to enable your energy healing ability except openness to the mystery of healing and a focused intent. Nevertheless, there is value to using techniques that healers from many cultures and over many centuries have developed and used successfully thousands, perhaps millions, of times. We are all insecure. It is somehow easier to trust in ancient techniques, symbols, and prayers than to simply trust in our self.

Healing intuitives, in many indigenous cultures, experience being healers without studying healing. They say they get their information directly from their spiritual guides during their personal meditations, dreams or vision quests. Fool's Crow, the Ceremonial Chief and Holy Man of the Teton Sioux said "Of course, it was not I who cured, it was the power from the outer world, and the visions and ceremonies had only made me like a hole through which the power could come to the two-leggeds. If I thought that I was doing it myself, the hole would close up and no power could come through."

Shamanic healing guides each have different strengths to support the power and intent of the shamanic healing practitioner. Healing guides might be animal totems, angels, goddesses or ancestral healers. Religious healers might draw on the help of prophets, saints or angels. Indigenous healers might draw on the help of animal totems, thunder beings or the four directions. Those in magical craft might draw down the moon or raise the power of Pan. Each person will be drawn to different types of healing guides that compliment and support their own background and needs. All are valid, all are a part of the great spiritual Source energy that is connected to the Great Creator. All efforts of this nature are part of our deliberate intent to work with spirit to co-create a positive transformation.

Anthropologists have studied shamans in places as diverse as Siberia, Indonesia, Africa, and Native America. (Note: In deference to the indigenous traditional shamans of the world and their ancient and unique tribal shamanic traditions, non-indigenous students of this form of healing art usually refer to themselves as shamanic practitioners or contemporary shamans.) The basic methods of journeying and healing have been found to be similar among these shamans. The Foundation for Shamanic Studies has synthesized the practices from various indigenous peoples into a certified program of shamanic study that has now trained thousands of shamanic practitioners in western countries. (Harner) (http://www.shamanism.org) The Foundation's efforts over the past three decades have now created many contemporary shamanic teachers, some of whom have also received training from specific indigenous shamans, such as those of Hawaii, the Andes, the Amazon and North American native tribes. Regardless of the pathway, the basic teachings and healing practices are similar.

For many, learning Reiki healing is more comfortable than shamanic healing. With Reiki, there is no journeying to upper or lower non-ordinary reality involved. It is done fully in the here and now, simply acting as a healing channel. Instead of relying on the active power and intent of the shamanic healing practitioner to travel into non-ordinary reality for support, the Reiki healing practitioner places trust that attunements from a Reiki Master will open a channel for universal healing energy. The Shaman tends to be more proactive than the Reiki healer and engages more aggressively in going after the disease and extracting it. Only the highest level of Reiki healers are so assertive in extraction techniques. Those healers that weave two or more techniques together have more tools to call on for healing, using what works best for the patient and his/her condition.

The practice of Reiki involves visualizing ancient Reiki healing symbols, said to be of Tibetan and Japanese origins, as a method of calling forth healing energy. These symbols were written in ancient Sanskrit texts in a Japanese monastery, undiscovered and not understood until Dr. Usui, a theologian schooled in Sanskrit, found them and went to a mountain for a lengthy meditation (essentially vision quest) to fully understand their healing nature.

The symbols that Usui Reiki and Karuna Reiki® Masters use to attune the healing hands of the Reiki students, i.e., for each level of skill (Level I, II, III and Master), are given in a step-by-step attunement rituals. Usually a student should practice for a period of 6-12 months between each level of attunement and training. The Reiki student learns that the correct, trustful use of these symbols will always open up the channel for spiritual healing energy to be directed to the place of disease in the patient's body. The healing energy typically passes through the center of the healer's palms.

In Reiki, the hands are placed on or slightly above the body at the chakra points and along the body extremities. The locations are shown in Reiki books. (Rand, Stein) Typically, the hands are slightly cupped, like a swimmers' hands, and the "bubbling well" in the center of the palm is slightly raised from touching the skin, while the thumb, digit finger, and three central fingers' tips are touching the body. Several levels of Reiki certification are available and many nurses, physical therapists and massage therapists are now certified in Reiki and use it in hospitals and physician offices. Some health insurance companies now cover

Reiki treatment, so long as it is prescribed by a physician and conducted by a certified Reiki practitioner. (http://www.reiki.org)

Healing Touch (also known as Therapeutic Touch) has similar side-by-side hand placements as Reiki but focuses on visualization of the energy flow and typically is done with the hands a small distance from the body. Also, in Healing Touch, there are other ways of placing hands at two separate chakra points, visualizing waves of energy between the hands, with the waves mixing like a whirlpool. Healing Touch was developed in the US several decades ago and has hundreds of thousands of health practitioners, with classes at medical schools for nurses and doctors. (http://www.healingtouch.net and http://www.healingtouchcanada.net for certified healing touch courses).

In the last decade, energy healing practitioners have developed special techniques for healing animals, such as Komitor Healing Touch for Animals® . Carol Komitor has extensively trained animal rescue workers, veterinarians, vet techs, and the owners of animals in the Healing Touch method (see http://healingtouch foranimals.com for certified healing touch for animals courses.)

Maintaining Focus on Manifesting the Healing Vision

More than 2,400 years ago, Lao Tzu, the great Chinese philosopher who created Taoism, wrote, "The highest good is like water. Water gives life to the ten thousand things and does not strive." Healing energy flows from the healer like an artesian well, continuous sustenance of clear, clean energy without worry about end... as it is channeled from our highest intent and limitless.

The energy-healing practitioner learns that regularly focusing on the intent of healing is enough to manifest a healing. The book entitled *Course in Miracles* is a series of meditations that teach how to detach from fear and anxiety that confuses and clutters your healing intent. When we become afraid, worrisome thoughts spin around and around in our heads, leaving no room for quiet clarity of good intent. The Course enables us to turn away from these fear-based thoughts.

Fools Crow, a famous Native American Sioux medicine man, described his healing efforts using song, prayer, drum and rattle and said, "The whole idea continues to be one of reaching freedom from fear."

The *Course in Miracles* teaches that there are two choices at every juncture in life..."fear or faith." The pre-requisite of healing is faith. The *Course* teaches that the greatest difficulty for the healer is to overcome doubts about the possibility of a healing. It is essential that the healer not allow continuation of symptoms to cause a lack of trust. After a healing treatment, the healer must believe that the healing will occur at a reasonable pace, as it should. Anything less is a lack of faith that can sabotage the healing effort.

Honoring Your Gift and Yourself

Healing practitioners often wonder whether this healing ability is so delicate, so mysterious, so fleeting, that they should do it without any compensation. In traditional Native American shamanic ways, people always bring a gift to the medicine person. First, the gift is an expression of giving permission for the healing. And, second, it is a way of honoring the life's work of the shaman to reach this point of healing ability. Bear Heart wrote: "Even if the medicine way I perform takes only a few minutes, how long did it take me to be able to do that in a few minutes? And what sacrifices were made on my part in order to learn it and earn the right to use it? By not giving something in exchange, a patient clearly disregards what it took for the medicine person to attain the knowledge that helped him."

I remember a story attributed to the famous artist, Picasso. As the story goes, Picasso was eating dinner and took the spine of the fish and all its attached bones and made an imprint within a plaster cast. When the cast dried, he painted a few strokes of color on it and placed it in a gallery for many thousands of dollars. His friend asked him how he could charge so much money for a work of art that took him only minutes to make. He replied, "It didn't take me minutes to do that work of art, it took my entire life." And so it is for each of us, each arriving at a healing event with a lifetime of experiences leading to that event and making it a bold work of art.

Honoring Your Guides

Most energy-healing practitioners honor their healing guides. For some Christians, the process involves lighting a candle in front of a cross, a statue of a saint or other holy entity. For some Balinese, flowers and foods are brought in a procession to a sacred tree or mountain. For some Native Americans, wooden or stone carvings are created of their animal guides, such as totem poles or dancing masks.

Native peoples often dance to honor their animal guides and increase their connectedness to them. During the dance they move like their animal guide might move, to the beat of a drum or rattle. Sometimes they will wear the fur pelt or feather of their animal guide or a carved mask. Dancing like this can help you connect spiritually to your guide, to feel its nature, to embrace its power. Bear has been my main totem. Aside from regular long naps, I often take a teaspoon of honey in honor of bear, while I look at a carving or picture of him. Yummy.

There is value to bringing your own creativity to the honoring process. It doesn't matter about your skill... only your intent. Three years ago, I wrote my first poem, in order to honor Wolf, who protected me when I was physically in danger and guided me during special healings requiring soul retrieval.

Wolf

Wolf, you do not come to me in journey,

Your energy, as my medicine way, I do not see.

I have not worn your fur nor danced to your rhythm.

But, in a threatened dark corner, when my being shifted,

It was your snarl that sounded as my lips lifted,

Your growl to which the cruel crowd parted.

Wolf, in the dark of night under the stars, I listen for your howl.

I feel your form moving in the shadows as you follow.

Bear is my guide in the world of the shaman.

He travels for my vision and brings healing medicine.

When my body is tired, battered and worn,

It is in his cave where I rest through cold and storm.

Jaguar is my magician in the world of lost souls.

Invisible, he transforms from mountain fawn to leopard spots

To panther black, to blend in the place of his retrieval roles.

But, Wolf, in the cool crisp night,

When my energy is alive and bright,

It is your tiered pack of nine I seek,

Your wolf language that I understand and speak.

Honoring Poem, by Sandra Cointreau, 2000

Chapter Five

Learning Animal Chakras, Meridians & Anatomy

The word chakra comes from the Sanskrit word for wheel or disk and refers to the spiraling energy vortex that exists at certain places in the energy fields of humans and animals. For most humans and animals, there are seven major chakra locations, and each of these corresponds to a specific set of organs focused around the adrenals (root chakra), ovaries/testes (sacral chakra), pancreas (solar plexus chakra), thymus (heart chakra), thyroid/parathyroid (throat chakra), pituitary (third eye chakra), and pineal (crown chakra). (Simpson)

Just as we have a physical spinal system, blood system, and nervous system that is an integrated network, we also have a non-physical energy system of chakras through which our life force flows and connects to universal source. Energy healing applies focused intention toward balancing the energy of the chakras and connecting the chakras as a balanced whole.

Human Chakra Points

The 7 human chakras are shown on the diagram below. Energy healing and balancing focuses first on these points. Humans are often are anxious about being touched, and energy healing normally begins with the chakras of the head, the crown and brow, and works toward the bottom chakra, the root.

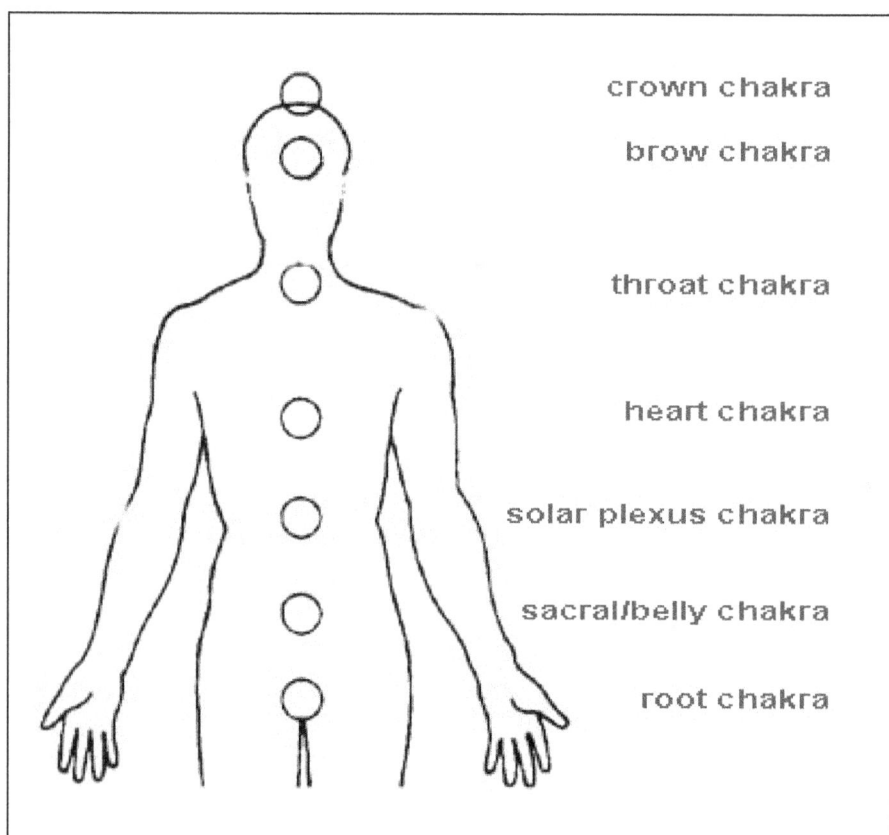

crown chakra

brow chakra

throat chakra

heart chakra

solar plexus chakra

sacral/belly chakra

root chakra

Animal Chakra Points

4-legged have the same 7 chakras as humans, except they are located according to animals' different anatomy and skeleton, as shown in the diagram below. The throat, solar plexus, sacral and root chakras are along the spine. The heart chakra is in the center of the animal chest, below the base of the neck.

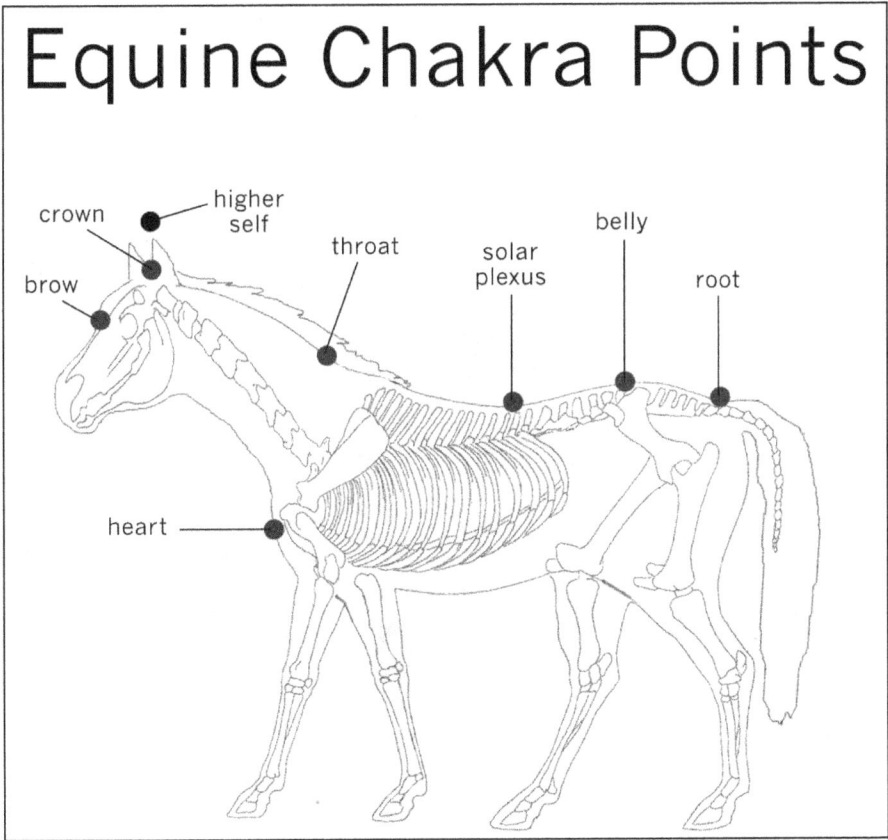

Equine Chakra Points

crown · higher self · brow · throat · solar plexus · belly · root · heart

The crown and brow chakra are near each other, the brow on the animal forehead and the crown at top of the head. When working with a small animal, like a miniature dog, the crown and brow points can be treated together, as the palm of the hand fits over both at one time.

Animal Skeletal and Organ Diagrams

For whatever animal you may be working on as an energy-healing practitioner, it is important to have knowledge of their general anatomy. Feline organs and the canine skeleton are shown on diagrams below.

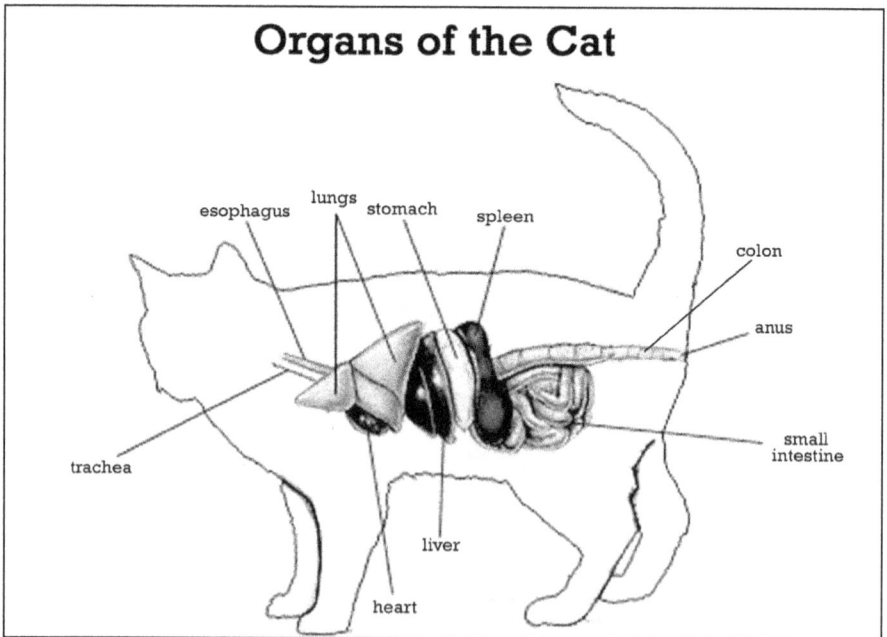

Organs of the Cat

esophagus · lungs · stomach · spleen · colon · anus · trachea · small intestine · liver · heart

Many animals lack the enzymes to break down cellulose in fibrous plants. Grazing animals rely on bacteria in special organs found uniquely in their digestive systems to enable foraging on fibrous plants. Horses, tapirs and rhinos have a cecum for that purpose, while cows, goats, deer, camels, hippos and sheep have a rumen. The horse system is not as efficient at extracting useable energy as the cow system, but it works faster and enables the horse to eat more relative to its size. Thus, the horse can survive on forage of poorer quality and live where ruminants might not be able to survive. (see Budiansky)

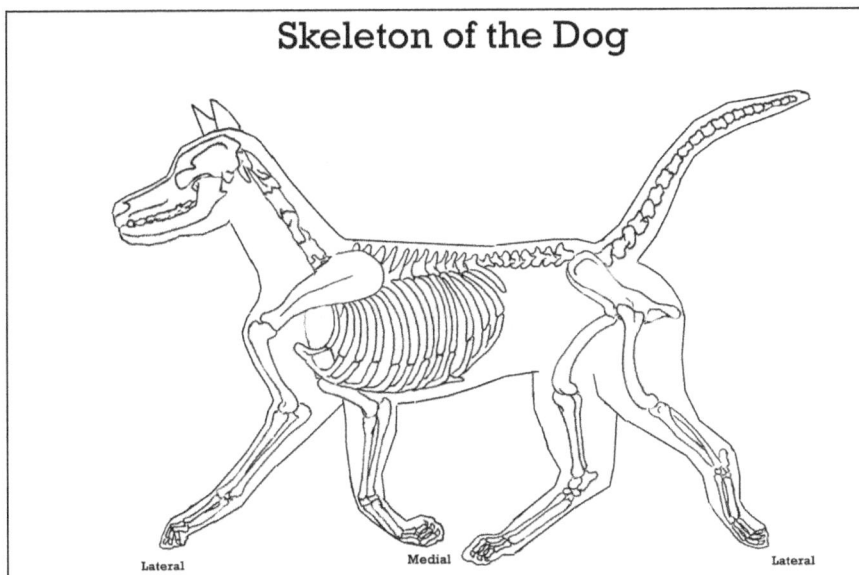

Skeleton of the Dog

Lateral   Medial   Lateral

Awareness of animal anatomy is important to any practitioner of energy healing. While only intent is needed to direct energy into the total body of the client, the energy-healing practitioner will feel areas of dis-ease (hot or dense areas) and knowing whether that matches the medical diagnosis is a useful affirmation.

Energy healing is non-invasive and needs only a feather-light touch, or no touch at all, to bring about healing. Energy healing is complementary to veterinary care, and is never harmful. After the chakras are balanced, the energy-healing practitioner may send energy directly at the part of the body diagnosed as injured or diseased.

Certified massage and acupressure therapists may directly apply pressure on or around specific muscles, joints, and organs. However, only expert therapists who are certified to conduct such therapy with the agreement of the animal's veterinarian should do this type of therapy.

Animal Meridians

Meridians and pressure points are energy pathways within the body. The key meridians and pressure points of four-legged animals are similar. Some of them are shown in the Annex of this Manual. For greater detail see detailed guides on acupressure for dogs and horses by Nancy Zidonis and Amy Snow. (For diagrams of canine meridians see the reference section at the end of this book.)

The meridians start at the base of the foot and move up the leg and along part of the body. There are special "ting" or "reflexology" points around the feet that may be given a small amount of physical pressure with one finger for a few seconds, if the animal feels comfortable about it.

If the diagnosis for the client shows a clear problem of a specific organ, it is useful to send energy healing along that meridian.

An Affirmation

When Rupert, the standard poodle puppy, had dangerously failing liver and kidney symptoms, energy was transmitted all along the meridians for the liver and kidney, and the designated final pressure points in the toe and paw were given a small squeeze of pressure, as in reflexology.

The urine bag started to fill up very quickly during this work, particularly when touching the kidney meridian's toe point.

Chapter Six

Learning Animal Behavior

Approaching the Animal

To be able to work with animals, it is helpful to understand them. They have feelings of affection, fear, stress, and happiness. They send out clear signals about their feelings and have a large repertoire of body, facial, and vocal language. They have even been shown to understand words in our vocal language, and respond to images. National Geographic (March, 2008) described a border collie of 6 years old that already had learned 341 English words, and a sheep that could recognize 50 faces.

Just as animals have thousands of years of genetic memory on how to migrate, hunt, forage, whelp their young, communicate, nest build, clean themselves, bury food for the next season, etc., you also have memories buried deep inside your ancient genetic coding. Humans have been living with domesticated dogs for over 10,000 years and with domesticated horses for over 7,500 years. (Coren, Budiansky) Given the potential for some genetic memory to exist in humans, it is reasonable to place some trust in your intuitive instincts about whether an animal is safe or not to approach.

Animals have a keen sense of smell and some believe that they can smell the odors that we excrete when we are afraid of them. Like many equestrians, I rub the scent of lavender into my wrists, so that this scent calms and balances the horse before I approach. There are also calming dried herbs, such as chamomile and yarrow, that can be offered to the horse by hand or in their feed, prior to a healing session. Dogs, cats, horses, and other animals will also respond to herbal tinctures or teas being added to their water. Both the scent and herbal medicinal properties of these chamomile and yarrow are calming. For discussion of calming herbs and aromas, see *Equine Herbals – An Earth Lodge Guide to Horse Wellness*. (Cointreau, http://www.earthlodgeherbals.com.) Also quite helpful for calming animals are flower essences, such as Bach's Flower Essences.

Knowing when it is Safe to Approach an Animal

A stressed animal can be a dangerous animal. Whether an animal is generally aggressive, or is defensive for the short term while feeling threatened, any stressed animal needs watchful and knowledgeable handling.

To recognize a stressed dog, the following signals might be given: hiding, shivering, whining, urinating, panting, hair standing up along the back, ears back, mouth corners turned down, eyes squinted, diarrhea, pulling away, down on the heels, lowered hips and tail, barring teeth and snarling, snapping.

To recognize a stressed horse, the following signals might be given: pawing the ground with one front hoof, head up high and eyes wide, ears back, backing away, head hung low and trembling, running the fence line, barring teeth, and attempting to bite.

Never go into the territorial space of an animal showing these signals until the animal has relaxed. Assume that the boundary of a stressed animal is at least 8 feet.

The energy-healing practitioner has an advantage over other healers, because she or he can work from some distance from the animal, if it appears unsafe to approach and lay on hands. While working from a distance, visualize the animal becoming relaxed and cooperating with the energy healing, and fully conduct the healing from a distance.

Sometimes, after working from a distance, the animal will approach the practitioner and gently place their body into the hands of the healer...often showing the healer precisely where they want to hands to be placed. Trust intuitively in the animal's knowing.

Learning Animal Communication

Animals communicate by their facial expressions, posture, movement, and sounds, as well as telepathically. Low-pitch sounds are usually used for

56

close-range communication among a pack, herd, flock, or pod, while high-pitch sounds are meant to carry across long distances to send out a signal of location, announce a danger, or call in the group.

Dogs have hundreds of body and sound expressions and there are detailed illustrations of many key expressions Stanley Coren's book called *How to Speak Dog* and David Mech's book *The Wolf: The Ecology and Behavior of an Endangered Species*. Coren provides detailed descriptions of the meanings of various barks, yelps, moans, and growls according to pitch, frequency, and volume. As a dog breeder, able to watch my family pack interact as adults and also with new litters of puppies, the Coren book provided detailed information to enable learning while watching. For the non-breeder, visiting your local off-leash dog park provides great opportunity to learn from multiple dogs interacting.

Animal behaviorists that studied groups of animals in the wild, as well as groups of animals in zoos or other captive situations wrote these books. It is noteworthy that most of the interactions among animals in groups are designed to achieve harmony. And while there are dominant animals, aggression is the exception rather than the norm.

Most animals prefer to be approached quietly and indirectly by strangers that have their head down and are slowing moving sideways toward the animal, with their eyes only glancing peripherally at them. A slight bending posture with shoulders sloped forward is less threatening than standing tall. Arms should be down and never raised quickly during the approach.

Wonderful training methods were developed for horses after observing wild horses interacting. (Roberts)

It is possible to mimic the animal's typical calming symbols. There are differences between calming signals and submission signals. Study and use the calming signals, particularly for aggressive animals. Only for a very submissive animal might submissive signals be useful. The following are two extreme examples used by dogs for submission and calming. Dogs under serious threat from a pack or particularly low ranking dogs will lie on their back to indicate total submission; while a high ranking dog will lie down on its stomach with its head up, Sphinx-like, to calm down a rowdy pack, or will sniff the ground as if finding an interesting scent, to distract a group posturing for a fight. (Rugaas) Yawning, without barring the teeth, is also a calming signal.

Dogs and horses live in groups and have distinct hierarchies within those groups. The lead (alpha) dogs and horses tend to be more "up" on their toes, heads held high, tails held up high (generally rigid, but sometimes trembling), and appear ever-vigilant to protect their groups and territories. Watch when two high-ranking dogs meet: the most alpha will be the dog on its toes and the last to wag its tail.

Cats are also territorial and have a hierarchy. Anyone with multiple cats in a barn or house quickly learns that the cat doors may be places of key contention, guarded by one to keep the others out. I find it best to create multiple cat openings so that all can reach shelter and food without stress.

An Affirmation

In 2002, I had an experience with a rescued horse named Doris, who had once been a champion jumper. She had a chronic leg injury, believed by the veterinarian to be a hairline fracture, and she had not been improving for months. She spent much of the day leaning against her shed wall, to take the weight off of the injured and painful leg. She was snarly and aggressive with most people and I was warned to be extremely careful around her. I admit that this mare scared me. She was the tallest of thoroughbreds and could move fast when she wanted, despite her injury. There had been a decision that she would be "put down" in a few days, so I felt a strong need to try to create a healing.

I stood near the shed door, ready to bolt over and out the door, and held out my hands sending her energy. After about 10 minutes, she walked over slowly and leaned her body into my hands. We stayed merged like that for about 30 minutes, and then she walked calmly away to return to her leaning wall.

The next day she was moving better, walking around her paddock. Later in the week, I returned for another session. This time, I asked the owner to hold her on a lead so that I could try to work my hands down her legs and feel safe that she would stand still. I was pleased at how well she could now stand on her own as I worked on her, and how receptive she seemed.

A few days later, I learned she was trotting in her paddock. I did one more healing session, this one was from outside the paddock, because she was very prancy and was displaying a rather intimidating bossy demeanor. Later that week, I learned that she jumped the fence and ran from her paddock. She was retrieved and lived well for several years more.

An in-depth Japanese study of a large group of several dozen cats sharing a open building and fenced area provides valuable insight to cat community dynamics. The females in the cattery would sometimes spat and flash claws at each other, but rarely duel. The toms regularly fought, but fights usually lasted less than 5 minutes and rarely resulted in any physical injury. The fighting established and maintained hierarchy. The Japanese study describes in detail the various dominant, aggressive, calming, and submission body postures of cats, including facial language of the eyes, ears and mouth. (Angel)

In horse herds, stallions will fight during mating season to obtain a herd of mares and the winning stallions will service a number of mares. While the battles look quite dramatic, injuries are normally modest and temporary. A beautiful book of photographs documenting the herd and mating behavior of horses is *Horses of the Carmargue*, with hundreds of photos of the wild horses living in the south of France. (Silvester)

It is noteworthy that there are fewer big injuries among the highest-ranking stallions, while the lowest ranking stallions fight more bitterly. (Budiansky) I've noticed with my family pack of poodles that the lowest ranking dogs fight more aggressively, since no one wants to be at the bottom of the ranking. Among wolves, social control by the highest ranking wolf is often asserted merely by a penetrating stare, while a middle ranking wolf is more likely to growl or bare its teeth to threats from lower ranking dogs trying to move up in the pack hierarchy. (Mech)

While humans are mostly familiar with the overt leadership behavior exhibited by alphas that are aggressive, there are horses and dogs that rule through more compassionate methods. A wonderful book describing horses that win over their herds without aggression is *Horses Never Lie: The Heart of Passive Leadership*. (Rashid) This book describes how one horse obtained a herd of mares without a direct confrontation, but through outsmarting the aggressive stallion into exhausting himself.

In dog and wolf packs, the alpha male will have a dedicated relationship with the most alpha female. Usually, only this alpha pair will mate, because of food and security needs of the pack. Typically, there is only one den that the pack can protect at a time and only one litter can occupy that den. The entire pack will bring food to feed the litter and even regurgitate partially digested food for the puppies when their mouths are licked. (Mech)

Trainers of horses and dogs often ignore the subtleties of pack and herd behavior and adopt the most obvious aggressive alpha behavior as their main mode of training. While trainers need to obtain the respect of the animals they are training, they also need to develop trust.

Based on in-depth study of wolves, harmonious packs with a tolerant alpha male will signal "active submission" by giving him a friendly greeting. However, with an aggressive alpha male, the lowest ranking wolves will greet him with "passive submission," which involves a complete demonstration of inferiority and helplessness. (Mech) These complex behaviors require study and observation to learn, but are worth studying to understand how to handle animals. L. David Mech's book *The Wolf* has many illustrative photos of wolves interacting with their packs in the wild to illustrate his detailed information on behavior.

According to Klaus Ferdinand Hempfling, to have an animal of the highest level of training, it is not a matter of trust or obedience, but "trust and obedience." Using many detailed photographs, Hempling's training book shows how to lead and lunge a horse, so that the trainer maintains the passive dominant position yet encourages trust. (Hempfling). Another recent horse-training manual encourages passive dominance, with significant observation effort and trust-building exercises. (Rashid)

Klaus Ferdinand Hempfling is considered one of the world's greatest trainers. He states that training for centuries of the great Spanish dancing stallions was done through "the transmission of our thoughts." (Hempfling)

For the energy-healing practitioner, where training is not an objective, the approach should always be gentle, passive and patient in building trust.

Animal Senses

Animals have different sensory perception from people. Horses have a large portion of their brain directed at controlling their movement. (Budiansky)

Dogs have extra brain capacity devoted to scent. Average-sized dogs have about 4 times more brain weight devoted to scent than humans have. (Coren)

Dogs are reported to be able to identify smells from 1,000 to 10,000 times better than humans. When working with animals, it is best to avoid smells that they don't like. Dogs appear to dislike citrus, citronella, spicy and pepper smells. (Coren) In my experience, most animals seem to be relaxed by the scent of lavender and have their appetite stimulated by the scent of mint. I always wear a flowery perfume when working with animals.

Dogs have their eyes more widely spaced than a person, which gives them greater peripheral vision. Horses' eyes are further spaced toward the sides of the head that gives them a nearly full 360-degree view, but with a narrow blind spot just in front of the nose and another behind the rump. While humans have a normal vision of 20/20, horses average about 20/33, dogs about 20/50, and cats about 20/100. Thus, a detail that humans can see at 20 feet away, horses see at 33 feet, dogs at 50 feet and cats at 100 feet. (Budiansky) Both horses and dogs in dim light see better than humans and are much more sensitive to movement.

Horses and dogs do not see color as humans see color. Humans have three kinds of cones to enable sensitivity to 3 different wavelengths of light, namely blue, green and orange, (respectively: short, medium and long wavelengths). (Coren) Horses are most able to discern red from gray, and to a lesser extent they can discern blue from gray. Green and gray appear the same to a horse. (Budiansky) Dogs seem most able to discern blue and yellow. Seeing yellow clearly implies that dogs have one cone that is configured to see a wavelength between that for green and orange. Dogs thus see the world as shades of blue, yellow and gray. Green, blue-green and orange are seen by dogs as gray and red is seen as nearly black. (Coren)

All animals make sounds that are designed to accomplish something important to the animal, and we are gradually learning to understand these. Horses, genetically surviving for millions of years within herds and on open grasslands, have evolved to be more dependent on visual rather than vocal communication. On the other hand, forest animals have developed with a wider vocal vocabulary. (Budiansky) Animals with larger ears typically can hear better and this also applies to dog breeds with larger ears, such as the Papillion.

Young humans with normal hearing can here up to 20,000 Hz, which is like the full range of a piano plus 28 more keys on the right hand side of a typical piano. Dogs can hear between 47,000 and 65,000 Hz, which is like adding 76 keys on the right hand side of a typical piano. Dogs with standing ears have an advantage over floppy eared dogs, as they are more able to rotate their ears to determine the direction that the sound is coming from. (Coren)

In dog verbal communication, it is often the pitch and frequency that mean more than the actual sound. And thus, most animals will be responsive to humans that make soft gentle nurturing sounds, regardless of the words spoken.

Approaching Animals

All animals have their own personal spatial needs, just as do people. Some animals can be approached to within a foot, while others have much larger spatial needs. Approach slowly and quietly. Unless you know the animal, I would recommend that you move sideways, stopping and waiting with each step, to slowly sense where the boundary of the animal's comfortable space is located. Ideally, walk slowly to the perceived boundary, based on observation of the animal's signs of comfort, and then wait for the animal to approach you.

In approaching both horses and dogs, it is calming to the animal if you turn your head away (sideways) and lower your eyes, looking only peripherally at the animal, until it appears to feel at ease. If it is not convenient to turn your head away, then simply turn your eyes to the side (Rugaas). Always avoid a frontal direct wide-eyed stare, or barring your teeth in a big smile, or raising your hand, as these signals may appear threatening and predatory.

Yawning is a calming signal for dogs, not a sign of fatigue or boredom. Dogs can be quickly put at ease if you look to the side and feign a large open-mouthed yawn. (Rugaas) Scratching or sniffing the ground indicates you have found something interesting and are not a threat to the dog... it will make a dog curious and forget her fear. With a frightened horse, look to the side, soften your shoulders, and lower your head.

An Affirmation

When I was in Iraq for 6 months as part of the first team doing infrastructure reconstruction in 2003, there was one very tense interaction where the marsh arab sheik that was taking me to a health clinic began having a disagreement with another sheik on whose land we were passing.

Dozens of men from both tribes gathered around their sheiks with me in the middle. There was lots of shouting, fists were held up, and vibrational sabers were rattling.

I lowered my shoulders, turned my head to the side, lowered my eyes, and began to scratch at the table in front of me as if there were something quite interesting to scratch. My Iraqi translator did the same, except for the scratching.

In time, the situation decompressed and everyone departed with warm smiles. Most westerners would have stared wide-eyed. I believe my study of animal behavior saved my life that day.

With horses, you may feign chewing by smacking your lips, which tells the horse that you are relaxed and grazing. (Roberts) Interestingly, according to the macaque literature at the National Zoo in Washington D.C., lip smacking is also a calming signal among monkeys and is said to be reassuring upon greeting and even appeasing after a squabble. Your intent with these actions is to make the animal feel that you are simply another creature in the pack, busy with your own activities, feeling safe and comfortable and not a threat to them.

As you get closer to your own animals, you may wish to bond with them in their own classic fashion – through grooming. This doesn't mean with clippers, scissors and brushes. But, in whatever way the animals groom those others in their packs that they like. Horses stand with their necks touching and each grooms the base of the neck of the other with their teeth. Dogs lick the eyes and ears of their friends. Monkeys scratch and pick each other's heads.

Handling the Animal Safely

Generally, I would not recommend doing hands-on energy healing of any animal type that you have no experience in handling. Also, it is always recommended that a familiar handler of the animal be present and that the handler have the animal on a lead.

If you are afraid, you should not handle an animal. It takes time to know how an animal moves and responds to your movements. Animals sense any fear and insecurity, and alpha animals may choose to show their dominance. You can easily do a healing from a safe distance outside of a dog fence or horse paddock, or beyond bird cage bars or a glass aquarium. A calm remote animal will be more receptive to your efforts than one that is stressed by close contact.

Certainly, it is common for animals to each have their personal spatial preferences, just as each of we human's do. Species is not the determinant. You will find cats that come to you for being petted and others that do not. Some dogs will allow their bellies to be rubbed, others will not. Some horses will let you touch their ears, others will not. Do not test their boundaries...respect them.

Approach slowly, and keep your body to the side with your head down, sidling up indirectly, rather than walking toward the animal directly. Lower your eyes and do not bear your teeth in a big smile, as this may look predatory to an animal. Keep your hands low, as raising them or gesturing them could be seen as aggressive.

Be relaxed and take your time, waiting between each movement to see how the animal is reacting. Stop if there is the slightest sign of tension, such as standing tall, head up, neck stretch, muscle tension, growling, stomping of front foot.

Step quietly and keep all the sounds in the area quiet during your approach. Speak in a low resonate voice, making soothing sounds or saying reassuring statements. Horses and dogs both use low pitched sounds and slow frequencies of sounds when close to each other to be friendly or when comforting their young, while they use high pitched sounds and fast frequencies of sounds to cry alarm or call the herd or pack together.

Read as much as you can from references such as those listed at the end of this manual. Visit shelters, kennels, stables, zoos, and other places where animals are available to be observed from a distance. Have trustworthy friends introduce you to their animals under safe conditions.

For energy healing, unless it is your own pet animal, dogs and horses should be on a lead, even if the lead is not being held. Cross-ties can also be used with a horse, connecting the horse securely to side walls in a barn, so that you can safely move around the horse to reach all body parts. If not available, ask the owner or an experienced handler to hold the lead. The lead needs to be on the animal so that it can be grabbed if needed, to restrain and calm the animal. The dog leash can be left hanging on the floor. However, the horse lead must be draped securely and tied over the neck, as horses frighten when they step on their lead.

With dogs, it is most comfortable and appropriate to place the dog on a grooming table in order to reach all body parts. If the dog will not stay on the table or stand while on the table, there are loops that go around the dog's body from a pole attached to the table. For a seriously ill dog that can only lie down, place it on the table or even on your lap, in order to work as best you can. If the dog is prone to biting, it is wise to use a soft mesh muzzle, sized to fit the dog.

Horses need to be on a loose lead, or on break-away crossties, because horses tend to resist if constrained. Ideally, an experienced person should be present, holding the halter and lead and talking softly to the horse, unless you know the horse well. If the horse is prone to biting, there are special muzzles for horses. However, if you are experienced in handling horses, you will already know how to position yourself to the side and behind the front shoulder to avoid biting.

For diagnosis, hold one hand on the horse to feel the front shoulder muscle. That muscle tenses and moves when the horse is about to move, and gives you a signal to move in the same direction to avoid having your feet crushed. The front shoulder also would tense if a horse were about to swing its head around to bite. With experience, your hand on the shoulder sends you an instant message to move in time to avoid injury.

As the horse settles down during an energy healing, you should still be mindful of where you place your feet. When bending down to treat the legs, always keep your head where it can't be kicked. When working the back legs, this means facing to the back and placing your body sideways to the horse; while your butt may be kicked, your head will be protected. When working on the front legs, face toward the front and place your body sideways to the horse; again, to place your head outside of the area of risk.

If you are not experienced with horses, it would be reasonable to wear a riding helmet during the healing activity.

An Affirmation

Based on my personal experience, I believe that animals are able to communicate with us from a distance. In the summer of 2003, I was living and working in Iraq when my old dog, Topaz, became ill. He had heart problems for years, but I received a telepathic message from him that he was dying, and knew this was something new.

When I called home, I was told by my daughter, Maya, that Topaz had been rapidly deteriorating physically for several weeks, and standing for hours in the small stream out back gazing at himself. His liver was failing along with his heart and he was, indeed, dying, and she had been trying to contact me for permission to have him euthanized.

Telepathic Animal Communication

Master horse trainers from Spain, called noble caballeros say "We ride our horses by the transmission of our thoughts, an art that has flourished here unchanged through the centuries." (Hempfling) Natural balance in riding is done through the will, manifesting one's vision, with a loose rein and no overt leg pressure. Subtle, virtually invisible signals are instinctively given, as the body aligns with our intent.

Professional animal communicators say that they receive visual images sent telepathically from animals and that they similarly send messages through telepathic images to their animal clients. To heal an animal, it is extremely important that the healing practitioner continuously envision the animal as improving in health and becoming well.

No matter what data comes from the medical testing or how negative the veterinarian opinion might be, it is essential to hold an image of healing... especially if it is true that the animal will see whatever vision you are thinking. This is perhaps the most difficult part of becoming a successful energy healing practitioner, i.e., your maintenance of a positive vision and your faith in your healing ability.

Energy healing practitioners at Level II learn to conduct distance healing. Many healers have had affirmations of animals responding well to healings. One energy-healing practitioner, Carol Komitor, works long distance with many competition horses, sending them healing images and boosting their confidence and composure for competitions. The results are significantly affirming and horse owners from all over the United States contact her for this help.

I did not actively begin my own animal communication studies and efforts until 2005, when I read and practiced the exercises in the book *Learning their Language*. While quite doubtful initially, I found the results impressively validating, as described in the box below.

Affirmations

I experienced the following affirmations with animals via phone calls with their owners, from hundreds of miles away:

* A dog owner called about her dog that was at the hospital and extremely ill. The veterinarians had not been able to determine the cause of the illness and the owner was wondering if it was time to put the dog down. I had a strong sense that the veterinarian would diagnosis the dog's illness shortly, and that it would be treatable. But, I had some strange communications and reluctantly said that the dog was more preoccupied with wanting to tell that he was upset about his red patterned couch being moved. The owner acknowledged that the red couch, which had been much loved for 9 years by the dog, had recently been thrown away. I then said that the dog was not comfortable to go on the new couches, which I was seeing as white. The owner agreed that the couches were white and the dog never went on them. We came up with a replacement dog bed in red and the dog was diagnosed in a few days, treated, and cured. The couch message provided a tangible affirmation that enabled trust in the other message.

* A dog owner called about her dog from a rescue shelter, wanting to know about its past experiences and how that was affecting its behavior. I received some information about how the dog had been treated, but had no way of knowing if the information was correct. Then the dog talked about its favorite blue velvet fabric. When I mentioned this, the owner affirmed that they owned a blue velvet couch that was a favorite of the dogs.

* A dog owner called about her dog at the hospital, wanting some help to see if there were any genetic possibilities that were the cause of the illness. As I had bred this dog, I knew that there weren't. I had a strong sense that the dog was having a severe immune response to a bacterial infection, such as Lyme or Leptospirosis, and suggested that the veterinarian explore that avenue. As usual, I didn't know if this was true to whether I might be projecting. Then, suddenly, I had the image of the dog being involved with a snake, and receiving a bite to the rear back leg. I mentioned this and the owner was very surprised. The dog had been bitten 3 months earlier, which provided some affirmation that the other message could be relevant.

* I was working in Afghanistan on the avian influenza outbreak containment. I had the opportunity to meet a dog adopted by one of the people living in Kabul while working on the reconstruction. I asked permission to sit in the garden with his dog and see if any communications came through. The dog told me it had reincarnated to be with this man and that it had been his puppy in a past life, when the man was eight years old. The dog told me it was not able to stay with this child and had been taken from him. I wondered how this Afghan dog had managed to become selected by the man. The man affirmed that the loss of that puppy had been a trauma in his childhood, and that he had chosen this dog because it reminded him of that puppy of thirty years ago. It had the same freckles on its snout!

Chapter Seven
Techniques of Diagnosis

Diagnosis by the healing intuitive is done using various techniques . As part of the diagnosis process, the healing practitioner relaxes the breath, clears the mind, and relinquishes judgment. According to the *A Course in Miracles*, intuition increases as we simply allow some of the limitations placed on our minds to be lifted. Don Juan Matus, the Mexican sorcerer who mentored Carlos Castaneda, stated, "The art of sorcerers is not really to choose, but to be subtle enough to acquiesce." Don Juan also taught "as you chase away your internal dialogue, other items of awareness begin to fill in the empty space."

Diagnosis using your Vision

Some healing intuitives rely on their vision of energy fields, including chakras and auras.

Such vision involves indirect sight or sight through the third eye (located in the center of the forehead). Barbara Ann Brennan in *Hands of Light* provides a number of exercises for enhancing the ability to see energy fields, or one's "clairvoyant vision." She suggests "gazing" at plants to see to see the greenish haze around their edge. Such a haze may look like heat waves or fog.

For most of us, gazing at a human body will enable us to see a grayish haze around the edge of the body. The more emotional or energized the person, the larger this haze will become, and our chances of seeing color will be optimized. One can practice seeing auras from the back of a church or auditorium, because ministers and performers tend to have their aura's greatly enhanced while they are speaking or otherwise performing. As people in the audience are emotionally affected, their auras will also increase, and you can watch this play out without feeling conspicuous.

Alan Watts, a scholar on Zen, in his *The Way of Zen* described that "non-action seeing" involved peripheral vision, wherein the eyes must be relaxed, not trying to see, and the mind must be calm, not grasping. He noted that *seeing* was often enhanced by darkness, where the energy lights can be most easily seen. To develop this talent can involve sitting in a darkened room and looking in a mirror with peripheral vision (as an artist would squint to see patterns in a complex landscape). Such squinting can be practiced with the eyes trying to see through the center of the forehead. Aside from trying to see one's aura in a mirror, it is possible to practice seeing energy between one's hands as they are placed about 6 inches apart. Practice pumping up your energy with your intent and seeing the energy field expanding.

Diagnosis using your Hands

Diagnosis also can be done through feeling differences in heat or energy through the palms of your hands. To sensitize your hands, you can hold your hands apart with the palms facing each other. Move the hand close, about 3 inches apart, and then further away, about 12 inches apart, and feel the energy field as you move your hands back and forth in this distance range of 3 inches to 12 inches. It should begin to feel like a pliable ball of energy that you can tangibly squeeze.

Do the same by using your palms to feel the energy field around your head, holding your palms about two feet from your ears and moving them back and forth from that outside distance to about 6 inches away. At some point, you will feel the edge of your energy field, and will be able to practice expanding that field by pumping out your energy.

After practicing and sensing your own energy, it is useful to work with others to feel their energy fields. They will be able to tell you if they feel you squeezing and pushing against their energy field. Some will have a large field around them, and others a very small field. Those experienced in managing their energy field will be able to pump theirs up so that you can feel and see it powerfully.

In the diagnosis of animals, the healing intuitive moves the palms of the hands from chakra to chakra (see table on page 78), along the center of the body, and then along the extremities, giving special attention to all key joints. Any change in perceived energy is noted. An area may feel hot, or the energy may feel dense or prickly. Particularly diseased areas are quite unpleasant to experience, and the healer may need to shake away the discomfort, or rinse their hands in cool water, after feeling the energy of a diseased area. Regardless of the information obtained from pet owners and medical diagnosis, always do your own scans. Sometimes, it will lead to important new information for the veterinarian, as the affirmation below discusses.

Affirmation

Friends brought me their large dog that could no longer walk. The dog had been diagnosed with hip dysplasia and was regularly being treated for pain. My friends carried the animal to my couch and I sat with it to do a body scan. Despite all they had told me about the hips, my hands kept going to the neck area above the shoulder blades. I told the dog's owners that I felt this area was the root of the dog's inability to walk, not the hips. The next day, they took the dog to a university veterinary clinic and had an MRI done. The dog had a cancer that had spread in a thin layer in the neck area, so there was no discernable lump that doctors had been able to feel in the neck. This cancer was compromising the dog's spine and nervous system. Surgery was conducted to remove the cancer and the dog recovered its ability to walk. While the dog's hip joint continued to have dysplasia that caused manageable pain, it was not the cause of its loss of ability to walk.

Chakra	Organ/System	Emotion	Color	Color Meaning
Crown	Pineal gland, upper brain, right eye	Feeling connected or separated	Violet, white	Spirit, truth
Brow	Pituitary gland, lower brain, left eye, ears, nose, nervous system	Trust or doubt	Indigo	Insight, clairvoyance
Throat	Thyroid gland, bronchial and vocal system, mouth, lungs, alimentary canal	Positive or negative attitudes	Blue	Sensitivity, communication
Heart	Thymus gland, heart, blood, vagus nerve, circulation immune system	Love or grief	Green or Pink	Healing, Emotion
Solar Plexus	Pancreas, liver, stomach, gall bladder, small intestine, skin nervous system	Anger or shame	Yellow	Intellect, gut intuition
Sacral	Gonad glands, reproductive system, spleen, circulation	Sorrow or guilt	Orange	Ambition
Root	Adrenal glands, spinal column, kidneys, large intestine, legs, feet, nose	Fear or anxiety	Red	Passion

Generally heat is readily detectable in areas where pain or inflammation is occurring. In areas of tumors or other types of disease, the healing intuitive may feel an area of heaviness or denseness. Some will sense an area of darkness. Each healing intuitive is different. The key is to note the areas where there are marked differences – whether differences in heat, color, light, emotion or density.

Diagnosis using a Pendulum

For many energy healers, precise location of a place of disease is done with a crystal pendulum. Typically, with human patients, the healer holds the pendulum over each chakra point and determines if it is moving, and in what direction. For animals that are not relaxed with a pendulum swinging over them, a different process should be used. For most animals, especially horses, the healer stands to the side of the animal, facing toward the tail end of the animal. One hand is placed on an animal chakra (or directed toward the animal chakra if the animal is unapproachable), while the other holds the pendulum. In this way, the animal does not see a spinning pendulum held over her, and may not even see it at all, given that the person is facing away from the animal's head.

The pendulum is ideally a good quality gemstone polished to a point and hung from a short string or chain of about 6 inches. The pendulum may move in a clockwise circle, a counter-clockwise circle, a straight swing, or not at all. Most healers experience that a clockwise circle denotes an open chakra and no movement or counterclockwise circles denote a blocked chakra. Blocked energy at any chakra is an indication of disease, either existing or pending. Distortion from a perfect circle, e.g., elliptical or straight-line movement, may indicate emotional distortion. If you wish to check how your pendulum is moving for you, simply ask it some very straightforward and predictable *yes* and *no* questions, then ask it which movement indicates an open chakra and which indicates a closed chakra.

The wider and faster the pendulum moves in a circle or swing, the more open the chakra. Energy healing of the chakras shows immediate results in opening the chakras. The pendulum is used both before and after energy healing for confirmation.

The use of crystals is an age-old method of divining, as described by Cherokee medicine man, Archie Same. The Cherokee used crystals to determine the cause of illness and also to learn how the healing should be done. The crystals were warmed and placed on the patient's body at different points and looked through until the healer could see the center of the illness. (Mails)

In distance healing, a pendulum can be used over a drawing of the client's body outline and chakras. It can also be used over a crystal glass filled with clear water that you can ask questions regarding each chakra and body part. You will get answers just as you would if the animal was right there with you.

For those working with shamanic techniques, it is common to journey to the heart-beat sound of a drum and obtain insightful information during the journey. For others, meditation can reveal a diagnosis and treatment path. It is also possible to ask for a dream that provides the needed information. In each of these cases, one articulates the desire for this outcome at the start of the journey, meditation or dream: clear, stated intent is always a powerful tool.

Chapter Eight

Protection & Clearing Before Healing

Preparation for healing is an important step toward focusing your healing intent. E. Barrie Kavasch taught me many of the space clearing and protection methods that I mention here, as she learned them from Native American elders in more than three decades of working closely with them for her books and teachings on native healing traditions, herbs, medicine wheel gardens, mound builders and ceremonies.

Clearing. Many energy-healing practitioners clear their spaces before healing. There is great value to first physically clearing the space by removing clutter, tidying all items, arranging furniture for an open flow, bringing in light and fresh air, sweeping and cleaning. To clear a space non-physically, you may use drumming, clapping or rattling of the perimeter and corners of rooms as these vibrations move and transform stagnant or negative energy lingering. You may use a bell, chime or singing bowl to vibrate the energy and raise the energy to a higher level.

Throughout India, the marvelous scent of Nag Champa comes from the incense burned for clearing and protecting temples, homes and offices. In North American tradition, sage, an antiseptic herb, is commonly burned and waved into all the spaces of the room by using a feather, fan or hand. Sage is considered a sacred herb by Native Americans for space clearing, as is flat cedar, sweetgrass and bearberry. In Central and South America, it is common to burn copal resin as part of shamanic rituals. Tobacco and corn meal are also considered sacred substances by Native Americans and can be sprinkled in and around an area to bless a space. Often, Native American shamanic rituals incorporate using sage, tobacco and corn in the four directions.

Catholics commonly use the Crucifix and Holy Water that has been blessed by a Priest for blessing a space and easing a deceased person through transition. Certain specially trained and gifted Priests will use these items and others to conduct exorcism rituals.

Many Orthodox Christians and Catholics burn frankincense to prepare the church for mass or an important ritual. In traditional Pagan or Magycal craft, incense is used to prepare the mind by stimulating genetic and past-life memories of affiliations with rituals done throughout history. It also is believed to draw spiritual entities through the traditional associations with these scents. It is believed to lift the energy of the intention to the ethos, in the same way that Native Americans believe that the smoke from their sacred prayer pipes carry their prayers to the heavens. Traditional witchcraft uses a wide range of incense types, depending on the type of magical process being conducted. (Grimassi)

Aside from space-clearing practices used by religions and indigenous peoples, there are also many techniques used in the ancient Chinese tradition of Feng Shui. (Hale, et. al.) Mirrors, fish tanks, and strategically-placed wind chimes are used to move energy in corners and around buildings. Certain plantings of trees and bushes can also be used in Feng Shui to redirect the flow of energy, such as the strong energy of traffic along a road or running water in a river. Stones can be strategically placed to hold energy in a home, especially in the directional spaces considered "money corners."

Healers typically clear themselves, as well as their spaces, in preparation for a healing. Some may fast or drink a cleansing herbal tea. Others may meditate and clear away all spin thoughts, bringing themselves fully into the moment. They also do ritual to ground themselves and connect to earth energy. Maya Cointreau has written a comprehensive book on methods of personal grounding and clearing, including a wide range of rituals, crystals, symbols, incense, oils, aromas, colors, herbs, and charms for personal grounding and clearing. (Cointreau)

Fool's Crow, a revered medicine man, was regarded by many to be the greatest Native American holy person of the last century. He spoke about being a hollow bone: "First I thought about all of the stumbling blocks about me that can get in *Wakan–Tanka's* (the native equivalent of Spirit or God) and the Helper's way when I want them to work in and through me. Then I asked them to remove these things so that I am a clean bone. They did this, and as I felt the obstacles coming out I grabbed them and threw them away. When all of this was done I felt fresh and clean. I saw myself as a hollow bone that is all shiny on the inside and empty.... I knew then that I was ready to serve *Wakan–Tanka* well, and I held up my hands to offer my thanksgiving and to tell Him how happy I was. Immediately, I could feel the power begin to come into me." (Mails)

The energy-healing practitioner, after clearing and preparing the healing space and their own personal vibration, places a clean cloth over the table to be used for the client and has clean colorful sheets or blankets to comfort the client and keep them warm during the healing. The healing practitioner often washes his or her hands just before the healing as part of the ritual to clear their own energy. Commonly, all rings and bracelets are removed, as these are believed to inhibit the flow of energy in the hands. A cool drink of water may be taken for clarity and grounding.

Often a candle is lit to symbolize the focus of intent of the energy healer. The candle may be placed with crystals or energy healing symbols, such as Reiki symbols, or simply with the client's name and an affirmation. A specific color of candle may be chosen to address the chakra needing special healing, such a pink, which brings love, yellow, which brings happiness, or red, which brings energy and passion.

With humans, the energy-healing practitioner typically stands behind the client with hands on the client's shoulders and asks permission to begin the healing process. With animals, it would be appropriate to telepathically image requesting permission to begin the healing process.

Protection

Reiki energy-healing practitioners typically write the Reiki symbols that they have learned onto their hands and over in the air over their clients. While these are healing symbols, they also provide protection relative to the intent of the healer, for both the healer and client. With a healer's advancement in Reiki, the number of attunement symbols received and available for use increases. While anyone can learn and write the Reiki symbols, most believe that only a Reiki Master is able to effectively give attunements to open a student's healing channels to a new vibrational healing level. Usui Reiki has three basic symbols, plus three symbols available only for Usui Reiki Master/Teacher healers. Karuna Reiki® has an additional eight basic symbols, plus an additional symbol only for Karuna Reiki® Master/Teacher healers.

Most healing practitioners say a prayer of protection and affirmation before healing, asking for the highest purest spiritual forces of the universe to protect the patient and the healer during the healing process. The requested outcome of any healing should be what is best for the highest purpose of the spiritual life of the patient.

Healing practitioners then hand themselves over to the process, as channels for universal spirit to work for the good of the patient.

My favorite protection prayer is the 91st Psalm, called "God our Protector," which I say aloud in each of the four directions during times that I feel the need for protection. A calligrapher created a beautiful written copy of this 91st Psalm for me, and it has hung in my home for more than 20 years.

Stones are often used for protection, especially when placed on window ledges of the home, and over doors. Some protective stones are also worn on a long cord over the solar plexus chakra. Carnelian stones are good for this purpose and can be tapped with a quartz crystal three times with purposeful intent in order to activate it. Jet absorbs negative energy, and is particularly useful for protection from psychic attack. (Grimassi) For Native Americans, the most commonly used stone of protection is turquoise, which is usually found in native necklaces, bracelets, rings and belts. Onyx, agate, citrine, jasper, topaz, ruby, and crystal quartz also are considered to be protective stones. Creating an image of a cross or a five- or six-pointed star, with any of these stones in the center would be a powerful ritual of protection.

In ancient magical craft, stones are used to mark a circle of protection around a ritual activity. In shamanic practice, stones also may be used, including stones to mark the four directions. Often the opening to the circle is placed facing the East, and a protective altar or symbol might be placed at the opening to guard the way.

During some shamanic rituals, there is a "gatekeeper" who remains stationed at the opening. If stones are not used, it is also possible to use cornmeal or salt to create protective boundary for a sacred circle.

Certain trees are believed to be protective, and you may wish to plant them on your property. Birch (*Betula alba*) and Hawthorn (*Craetegus oxyacantha*) are believed to be empowered to ward off evil. Hazel (*Ilex aquifolium*) is associated with protection. (Grimassi)

To protect your home or work space, you may use salt laid down along the boundaries as you walk the perimeter and say aloud a prayer of protection. I also like to tie red ribbons in the trees and bushes around my property to affirm my intent that it is a sacred and safe place. In some Asian countries, people sprinkle salt or white lime powder outside their front door each day and write protective words with the white salt or lime.

There are a number of protection images you may choose among. Some wear an image of their totem, or power animal; or place fur or feathers from their totem in their home. Some northwestern tribes in Canada and the US create wooden totem poles, showing images of their power animals. Native Americans also create Medicine Shields, which depict spiritual symbols of protection specific to them and their guides. These shields may include the hide, bone, fur, or feather of their totem. Others erect a statue of a saint, such as St. Teresa or St. Germaine; or sacred deity, such as Kuan Yin, one of the four great Bodhisattvas in Buddhism and the Chinese Goddess of compassion and mercy. In some eastern Mediterranean countries, such as Turkey and

Cyprus, many people wear a glass image of an eye, or place the glass eye near their doors and windows, to protect themselves from "the evil eye," the negative energy of psychic attack.

When we have a problem, we tend to focus on a solution. Many of us take our vitamins only when we are not feeling in optimum health and do our protection only when feeling some negative energy or having that gut feeling that there is something wrong. But the old adage "Prevention is the best cure" applies here and protection should be a routine part of energy healing work.

We tend to forget about grounding as part of protection. Many healers and patients love the feeling of expansiveness that this work involves and forget to put their energy back and ground themselves. This can lead to accidents by not being fully in the body. After healing, it is appropriate to ask your patient to visualize their legs as hollow tubes and have them exhale down through those tubes into the earth and fully connect with the earth. Sometimes, placing your hands on the tops of their feet while they stand and breathe into the earth is very helpful to their effort as you send energy down through their feet and root them to the earth's grounding and protective energy. For an informative and inspired book on a wide range of techniques see Maya Cointreau's book, *Grounding and Clearing: Being Present in the New Age.*

In traditional magycal craft, the pentagram is the most common protective symbol used. This five-pointed star symbolizes the four elements of earth, air, fire and water, with Spirit, also called Ether, God or Source, being the overlighting fifth element. The fifth element is the energy that connects all living things. The pentacle was traditionally placed on shields of warriors and was on the shield of the legendary knight Gawain. In the Craft, it is sometimes placed at the opening of a ritual circle. (Grimassi) Its protective symbolism was so widely appreciated that it became the symbol used by US Sheriff's as their badge and homeowners throughout the US hang the five pointed start on the sides of their homes and barns. Wearing a pentagram is believed by some to provide powerful protection, with pentagrams on rings and amulets being most common. For a protective ritual, large pentagrams can be traced in the air around you, in the four directions, and noting the 5 elements embraced by the pentagram's symbology.

Some healers visualize being surrounded by a circle of white light and they use their hands to draw an encircling bubble of light around themselves. If they have been experiencing negativity being sent to them, they may also see that bubble of white light as impenetrable and reflective, bouncing any negative back to the sender. Others use white pyramids of light and see themselves safely within.

While it is karmically inappropriate, energetically unethical and potentially dangerous to send negative energy to someone, cast a negative spell or restrict a person's free will, there is no karmic issue in simply reflecting or deflecting negative energy without any contact, curiosity, attachment or anger. It is energetically always true that "what goes around comes around," so never seek revenge. With complete emotional detachment, one can hand over justice to the Great Creator and avoid any karmic involvement. One tool for handing over justice is reading aloud the 94th Psalm, called "God the Judge of All," in the four directions.

On a daily basis, you may wish to do a simple protective ritual as part of your morning shower or daily meditation. Every day, I anoint myself on my "psychic door," the back of the neck at the base of the head, to keep away negative spiritual energy and any intrusion from low-level spirits. For this anointing, the most important component is my intention of protection. You may use holy water. I use a lovely mixture of sea salt, almond oil, and spikenard essential oil, and rub it into the back of my neck. Spikenard is the anointing oil that was used by Mary Magdalene on Jesus Christ and its scent seems to bring up ancient genetic memories and associations of protection and compassion. The sea salt is a cleanser and its rough texture feels good as I massage it into my typically-tense neck area. Then I put the mixture on areas of my body where I may have accumulated some negative energy from my healing work, perhaps by inadvertently picking it up from a client. During this ritual, you may wish to use your finger to trace a symbol on yourself, such as the cross or the pentagram.

Native Americans may lie on the earth and call in the healing protective power of Mother Earth, releasing any tension and accumulated stress into the earth who so kindly receives and dissipates it. After such a ritual, it would be appropriate to offer to the Earth the tobacco and corn in the four directions, in gratitude and respect.

Just as you begin your healing ritual, you may think or state your own prayer of protection and affirmation. Here is a sample prayer:

A Prayer to Protect the Energy Healing

Today, I hand myself over to the Universal Mysteries,
without Ego or Attachment to my desires.
Today, I hand myself over to the Universal Mysteries,
and trust in the support that will come for my good intent.
Today, I focus my intent on this healing,
and know that this healing will succeed and be lasting.
Fear and worry have no place in my heart,
because it is filled with compassion.

My client and I are fully protected by my good intent,

and only the highest good comes from this healing.

Chapter Nine

The Basics of Energy Healing & Balancing

There are many techniques for healing. But, all the ten thousand things are one, and all these techniques are fundamentally the same. The basic ingredients are:

- Calm yourself and connect compassionately to your client.

- Relax your breathing.

- Connect to earth energy through your feet and your exhaled breath.

- Allow earth energy to ground and protect you.

- Feel your center and be at peace with yourself.

- State your intention, write your intention, or see yourself writing your intention on a blank screen in your mind.

- Close with a statement showing your faith in the healing, such as "this is my desire, so shall it be."

- Prepare your hands for the energy healing, which may include writing healing symbols into your hands (reiki symbols) or protective symbols (pentagon, cross, ohm symbol) or visualizing colored healing light coming from your hands.

- See the healing moving and successfully performing a healing, see every cell being affected and all cells working together to find perfect harmony and well being together, see every cell shifting toward a complete healing that begins instantly and see every cell maintaining this shift until the healing is complete.

- Sense the connection between you and the being that you are helping to heal and enjoy with love the sensation of the healing,

- Express gratitude for the beautiful experience and your affirmation of the success of this healing.

- State your firm belief that the healing is complete and successful.

Energy healing for animals, including critters as diverse as fish, birds, reptiles and insects, is essentially the same as energy healing for people. Plants, like animals and humans, also will respond to energy healing. Plants are well documented as responding to compassionate thoughts and words. If you work energetically on one plant in your home, don't be surprised if all the other plants also improve. Plants apparently communicate with each other and register the pain and pleasure of each other as a coherent community. Likewise, it has been shown that harming a single plant in your garden, will likely lead to measurable stress on monitoring devices of neighboring plants. (McTaggert, *The Intention Experiment*) While the community impact on animals is not so well documented as it is on plants, I expect that four-legged beings in a herd, winged beings in a flock or swarm, and finned beings in a pod will be sufficiently connected energetically that pain to one will stress the others, possibly even cause pain to the others.

Seemingly inert objects also can be positively affected by energy healing. Equipment can respond to your energy surges, even if they are not negative. I have been learning the hard way to avoid strong moods around my laptops, having crashed three laptops in the past 5 years. Many people involved in psychic activities and energy healing experience problems with watches, computers and other equipment, because their energy fields are strong and when their moods shift they may unwittingly disrupt equipment. This phenomenon was reported by Lynn McTaggert in her brilliant book, called "The Intention Experiment," that elegantly weaves together new scientific understandings about quantum physics with documentation from numerous experiments testing "intention." This book is a "must read" for all energy healers wanting to understand the scientific basis for their abilities and to have the affirmations that come from deeper understanding and documented proofs.

All beings have intelligence, feelings, fears and desires, and all are amenable to the same non-physical communication and healing as humans. Whether the being is two-legged, four-legged, winged or finned, the healing concepts are the same for each. Organs, chakras, meridians, etc, might be located differently in each, but the energy healer does not need to be concerned about these specifics.

Your healing will go where it needs to go, as Source will be directing it to the right place. You have only to hand it over to Source. Based on their background, some beings might best be healed without physical contact. When in doubt about any animal, simply focus on healing remotely. For remote healing, there is no difference needed in the methods from one type of being to another, whether it is a fish, bird, reptile, animal or human. For step-by-step guidance used by Reiki practitioners for different kinds of species, including birds and fish, see the book *Animal Reiki* by Elizabeth Fulton and Kathleen Prasad with its many illustrative photos of hand placement on various animals.

As discussed in the opening chapter of this book, energy healing of all beings is rooted in intention. The intention may be positively framed, by asking and by clearly intending and faithfully expecting, for the total well-being of the animal or person to be healed, i.e., for the return to the most perfect natural state of that being. The intention may be negatively framed, by asking and clearly intending, for the total destruction of invasive cells, toxins, or disease symptoms that appear to be inhibiting well-being.

Some healers (and ill people doing self-healing) like to envision themselves doing battle with disease and overcoming the disease. However, this can backfire if it involves aggressive emotion or feelings of anger or competition with the disease, since negative emotion often brings new disease. Also, strong focus on battling the disease may even over-stimulate body responses, such as immune responses, that may start destroying healthy cells after finishing with the unhealthy cells. Focusing on the diseased aspects of a patient may actually bring elements of the illness to the healer. Yes, I've been there and done that, and it's not a great experience. Taking on diseased aspects of others can be very hard for you to heal in yourself, as the intention involved in taking it on was very intense and your desire to heal your own body might not be as intense. Unless we are very clear and successfully experienced in manifesting our healing intentions, I recommend that we frame the request for healing in a positive manner and thenrelax in a feeling of happy expectation of pleasant and steady progression to perfect well-being.

Determining our intention is therefore the first step in healing. An accomplished healer with a well developed connection to their inner being, higher power and Source, may simply state the intention of perfect well-being for someone and hand it over to make it happen. For thousands of years, there have been many reported demonstrations of famous healers walking up to infirmed people, touching them and saying "you are healed," and witnesses observing an instantaneous healing. Such a healer doesn't need to know any details about the being to heal the malady. Many experiences and affirmations have enabled the accomplished healer to simply "intend" it and let it go.

For those of us that need to work a bit more to feel more confident in the manifestation of our healing intentions, the following activities are useful in developing and expressing your intention.

- A supportive step in developing your healing intention is to satisfy your mind. Many of us need to understand a situation before we proceed. If you are an information-junky and enjoy being precise, don't deny your natural tendencies. Feed the mind enough to feel satisfied and confident that you can articulate a successful healing intention. To do this, calmly and comfortably interact with the being to be healed and talk to the being's person, if there is one, to understand what has been happening with the being. Learn about the symptoms of the injury or disease, understand what medical people have determined or tried to determine, and enjoy the compassionate bonding of relaxing with the being and its person. If the being is a wild animal, you may need to rely on your personal observations and intuition to understand the situation.

- After a period of interacting and observing, sit with the being and allow your intuition to quietly connect. Animal communication is very vivid, showing visual images with textures during the communication, and also passing thoughts that are automatically translated into words that are clear and precise. Animal communication often enables you to feel that you are actually in an animal's scene feeling the light and temperature of the scene, seeing the animal in its normal scene as if you are there out of your body and in another time zone, and hearing what is happening in the scene with any other beings in the scene. Let the thoughts come and intuitively experience the message that the animal wants to convey about why it is feeling the way that it is feeling.

- Communication with an animal is not limited to understanding the cause and nature of an illness. It can and should involve understanding what the animal wants as an outcome, recognizing that perfect health is not always what is wanted. Animals, like people, sometimes have and hold on to an infirmity as a form of protection for something they don't want to deal with or as a way of obtaining attention. To animals and people, negative attention is often preferred to no attention at all. In the dialogue with the animal, if this is understood to be the case and the illness is not life-threatening, it might be appropriate to postpone the healing and interact with the animal's person to address the animal's concerns about their relationship and any positive aspects from having an infirmity that the animal may be feeling.

- Once you have done enough communications, observations, intuitive perceptions and building of rapport and compassion that you are satisfied, begin the process to develop your intention. Contemplate the information and feelings that you have developed. If it helps you in clarifying your thoughts, write down some notes about the information and feelings you have developed. Then, you may simply express your intention mentally, or you may write it down. In ancient magyckal craft and present day Wicca, an intention may be written as an affirmation structured in poetic form. For others, an intention may be written as a prayer honoring their beloved Source entity and asking for help, as is done in the Christian "Lord's Prayer" or Catholic "Hail Mary." In some cultures and times, an intention is done in a humble begging manner. When taken to an extreme, this begging may be accompanied by an act of suffering, whether fasting, promising to give up something important in the future if the intention is granted, walking along a pilgrimage path, crawling to an altar, lashing with a whip, wearing a hair shirt, or tying skin to a pole during a Sun Dance. As an energy healer, you need to believe in your intention having the potential for success, and thus you need to work within your traditional belief systems. If that means a spell or a prayer, so be it. All these ten thousand things are one…they are all expressed intentions.

- If you have decided to write down your intention, you may find it helpful to place it on some type of altar where you leave it with a candle or incense to carry the intention upon the smoke to the heavens. Whether you have written or simply mentally articulated your intention, you now need to decide whether you are done, or whether you feel the need to go through additional healing processes.

Articulating an intention and focusing over time on its manifestation may be enough for it to be realized. But, there may be an intense experience for the client in receiving an energy healing process, such as a reiki or shamanic healing. If we decide to proceed through an energy healing process, the energy that we send during the process will be felt by the being that is receiving it. It may feel like a gentle cool wave of cleansing energy washing over their body, focused extraction of pain or illness, or soothing warmth. It may feel like being drawn into another space or reality, as if transported from the discomfort of the body and being weightlessly floating. With people, it is fascinating to obtain their verbal feedback. Many will say they felt some of the above cool, warm or transporting sensations. Others will talk about experiencing the wings of birds fluttering inside their body clearing out disease, multiple hands being laid on their body at once, a sensation of colored light moving through their body, the disassembling of all their cells and reconfiguration, their various life experiences being healed retroactively, or even past life experiences being healed. We won't know how the animal feels on the receiving end of energy healing, unless we are able to receive the information intuitively when they tell us. With animals, if they are in our presence, they will tell us with their facial expression of gratitude and their body language. The experiences, while uniformly wonderful, are as diverse as the beings themselves.

A highly evolved energy healer may be in a state of readiness to do healing without any preparation. They are so routinely in and out of the healing state that in a sudden emergency, they can snap into that state in an instant. For most of us, there are procedures that enable and maximize our healing ability. In my experience, these activities can support our powering up of intention and transmittal of energy healing:

- Choose a space where you are comfortable and at peace. Having a special place for healing will create a positive residual energy field that accumulates in that space, making it sacred and safe for you to quickly transition into a healing mode. Color, crystals, art, music, aromas, plants and animal friends that make you feel positive and at one with your inner being, your higher self, are recommended. If you are not able to create a physical space for your effort, you may create a non-physical space in non-ordinary reality. For some, this may be a space in the lower world populated with your animal totems and traditional healing guides. You may meet them in an earth lodge, a crystal cave, a beautiful meadow, a dessert plateau, or an underwater cavern. For others, this may be in a space in the upper world populated with celestial healing guides. You may meet them in a beautiful temple in the sky or on the surface of a cloud. Over time, as you meditate to create you own inner sacred spaces for healing; you will be able to get to them quickly and easily and obtain all the support from your helpers that you desire.

- Before beginning your healing activity, prepare your physical space to be quiet and without disturbance or disruption for the time you plan to be conducting the healing. This may involve disconnecting the phone, closing doors, asking people to leave you undisturbed for a period of time.

- Start by sitting comfortably in a balanced body position that is open to your energy flow. Avoid crossing your legs or arms or being lopsided.

- Slow your breathing and concentrate on breathing slowly into your belly and extending your belly as you fill your torso with fresh air. As you breathe out, exhale long and slow, fully emptying your torso of air. Do this until you are completely in the moment with your breath and your body's action of breathing.

- If the sacred space that you have created is non-physical, rather than a physical healing space that you have physically created, this is the time to journey to that space. For many, journey to that non-physical healing space is most easily done if there is repetitive sound, such as heart-beat drumming, rattling, singing bowls, or chanting. The more you journey to this space, the less help you will need. In time, you may get there in an instant by simply intending to be there.

- In this state of being in the present and in your sacred healing space, contemplate your intention, stating it to your physical self and to your inner self, and requesting it of your higher self and to all the connections to Source or the Great Creator that you believe in, whether they are yogis, saints, angels, guides or totems, and/or to Source itself. Repeat your intention and continue to breathe fully and comfortably for some time with your repetition of your intention in your mind.

- Begin to breathe your intention into and through your healing hands, visioning your arms as hollow tubes and imaging your breath carrying your healing energy through your arms and out through the center of your palm to the being receiving the healing. With each in breath, hold in your mind your intention, and with each exhale breath that intention of healing out.

- Continue to do this breathing of outward healing until you intuitively feel that you have completed the process. With animals, if they move away from you or become restless, they are signaling that they have had enough for the moment. Infant animals, like babies, should normally have smaller doses of healing, so be very responsive to their signals. Adult animals generally need longer doses of healing but will also signal when they have had enough. However, animals that are extremely ill or dying may need a series of long multi-hour sessions.

Many human beings receiving healing are so into their minds and the worries that are spinning around. Their thoughts can put them easily back into illness and sabotage a healing. Therefore, I like for them to be involved in their own healing. In my work as a sanitary engineer for nearly forty years, I have always tried to teach my clients in developing countries how to fish, not just to give them fish. If the client is willing, I follow the same practice. And so, as part of a healing, I will show them how to use their breathing to stop their thoughts and cleanse their chakras and cells, and I will put a healing attunement into their hands so that they can do their own self healing.

My steps for instructing my clients in the cleansing breath are as follows:

- Sit comfortably in a chair with your back straight, shoulders comfortable, arms relaxed, hands relaxed in your lap, and legs comfortably placed with feet on the floor. Keep your legs and arms uncrossed and open, to enable good energy flow throughout the time needed to do this cleansing breath and self healing, which could last 20 – 30 minutes.

- Breathe three times deeply into your belly and then out, or longer if it takes that to begin to get your breath steady and long. Inhaling deeply is important. Exhaling fully is important to healing and enables more fresh air to come in, so give the most attention to exhaling every bit of air you can.

- Breathe three times deeply, visualizing that the breath is leaving you through your crown chakra. Then do the same through each chakra, breathing out three times through your third eye chakra, throat chakra, heart chakra, solar plexus chakra, sacral chakra and root chakra.

- Breathe three times deeply, breathing out through the arms and shooting the breath out through the palms. Then do the same through the legs, shooting the breath out though the arches of your feet.

- Breathe for the next 10-20 minutes deeply, breathing out through all the pores of your skin, as if your skin is completely permeable. Visualize that every cell in your body is receiving fresh air and that every cell is releasing any toxins, disease, or tension. Each cell is opening fully and feeling joy and healing. As you do this, you will feel your energy field expanding and you will merge fully with your surroundings, losing all sense of boundaries of your physical body.

- When done, breathe down through your legs and connect to earth energy through your feet, until you feel fully grounded.

The next step is to put a healing intention into the client's hands, so that they can do self-healing, or so that they can regularly contribute to the healing of themselves and their animals. Animals are so in tune with their persons, often taking the stress and disease of their persons. Any support to the animal's caregiver is helpful in the long-term well-being of the animal.

If you are a reiki master, you may draw the some of the basic reiki symbols into your client's hands and give the reiki attunement ritual for the level of reiki that you intuitively believe will best help and support your client's needs. Explain to the client that this self-healing attunement is not certified training and does not enable them to go into practice.

To put a non-reiki healing intention into the client's hands, simply as placing your finger tips on their hands as you intend in your mind that their healing channels open. Ask your guides or your inner being to help you with this intention. You may wish to write the word "healer" in your client's hands, or write the "ohm" sign that universally indicates compassion and well-being. Keep your fingertips on the palms of your client's hands until you feel that you have opened the channels. If you intend it strongly, their healing channels will open and they will be able to experience the flow of healing energy through themselves.

As a next step, I lead the client through an experience of self-healing. To do this, I would instruct with the following steps after the cleansing breath and attunement.

- Breathe slowing in and out until feeling that the breath is even and deep.

- Breathe deeply into the belly and then breathe out through the arms and palms of the hands, feeling the energy flow on the exhale, intending the healing energy to be passing out with the breath through the palms of the hands.

- Continue this method of breathing out through the arms and palms of the hands, placing the hands on various chakras and parts of the body to experience the healing energy.

- Continue this process until the client feels that they are able to sense the warmth and energetic flow into their bodies from this self-healing process. They will then understand how to do this process for their animals, as maintenance and support for any healing work that you have done for their animals.

- Finish with the client breathing out through the legs and into the earth to become fully grounded.

Shamanic Techniques

Shamanic journeying is quite different from the above steps of energy healing. Once the intention is developed and body is put into a comfortable position with relaxed breathing, the shamanic practitioner journeys to an inner world connecting to the inner/higher self and travels in non-ordinary reality to solicit totems and guides to conduct a healing. Once the journey starts, the shamanic practitioner does not direct the process. The various totems and guides literally whisk the shaman away. No matter where or how the journey starts, the path of the journey will be up to the totems and guides. The potential experiences are too diverse and amazing to explain.

A journey may be as simple as receiving an intuitive feeling of connection to one's inner/higher self and requesting soul support for one's intention. In such a case, there may be no visual images or sounds. But, for most, a journey involves meeting one's guides in a visually interesting sacred space in non-ordinary reality, honoring them in some ritual way, asking them to conduct the healing, thanking them, and then returning. This type of journey will involve sound, texture and a sensation of psychically being in that other reality. The different non-ordinary realities encountered may be a complicated as traveling into a dark mysterious wetland of lost soul pieces and encountering a series of frightening obstacles and entities to retrieve those soul pieces. It may be as simple and sweet as walking in a meadow where one encounters gentle animal totems like rabbit and deer. It may be as liberating as flying up above the clouds to a celestial world of beautiful temples and monastic guides and angels. Hank Wesselman provides a concise overview of instructions for beginners in his book *Spirit Medicine: Healing in the Sacred Realms*, which is accompanied by a CD containing drumming and rattling to facilitate the journey process.

A detailed explanation of soul retrieval, an advanced level of shamanic work, is covered in the book *Soul Retrieval: Mending the Fragmented Self* by Sandra Ingerman. The concepts of soul loss are explained. Upon return from soul retrieval, the recovered soul pieces are then physically returned to the being, sometimes by blowing them into the chest with a "shaman's breath," sometimes by handing them over encapsulated in a crystal for the being to hold and allow the soul pieces to be absorbed. Ingerman's book gives a number of examples of journeys from her extensive practice and teaching experiences.

Dreaming, Meditating or Journeying for Insight

There are various ways of obtaining insight on a client's needs. Meditating allows insight to float to the surface within minutes of quieting the mind. The important step is to write down the request for insight before starting, then fully detach and empty the mind of all thoughts.   Gradually intuitive insights will arise. Dismiss each one as it comes, never dwelling on it, and allow the next one to arise and so on.

Ask for a dream that provides insight.   The dreamer writes down the request for a dream just before going to sleep and leaves a paper and pen ready to record the answer to the request upon awakening.   In the dream world, different people, animals and objects will be symbols of the message that your intuition is trying to provide.   You may be conscious of dreaming, even though you are asleep.   When nearing the end of the dream, ease out of it slowly and prepare to write it down as soon as you awaken.   By moving very slowly and staying partly in a prone position, the brain blood flow will stay slow and the dream will not be erased.   Simply roll toward your paper and pen very carefully and jot down enough thoughts so that the dream will be remembered.   Alarm clocks ringing, lights turning on, dogs or children jumping on the bed, all have a strong dream-eraser impact.

With recordings of drumming or rattling, or by doing your own drumming or rattling, it is possible to induce a relaxed state that enables journeying.   Journeying is travel in non-ordinary reality, while fully awake.   It is similar to dreaming, but much more vivid and typically it is a continuous sequence of events whose meanings are clearer.

Michael Harner studied various indigenous peoples around the world and developed a technique that works for most students of shamanism.   Harner's Foundation for Shamanic Studies has recordings available to enable journeying by drumming or Tibetan bowl. Hank Wesselman has developed a delightful guide for journeying and a pleasant sounding drumming CD comes with his guidebook.

During journeying, it is possible to ask for guidance on healings, and also for help with healings. Very explicit visions appear and there are usually affirmations from the client that the journey has had a positive impact. Not everyone is suited for journeying but, for those that take the plunge, the experiences can be amazing.

The Value of Good Training

Whether the healing process chosen is healing touch, reiki, chi kung, or shamanic journey, specialized training in that technique is recommended. Training in groups is particular worthy, as the energy of the group supports the entire learning experience and the group members share their experiences. Also, during group training, there is opportunity to work on each other and experience the difference in the healing hands of different members of the group, as well as obtain feedback from others on how your own efforts felt to them. The better the teacher...the better the training.

Many healers know several techniques of healing, particularly both reiki and shamanism, and intuitively choose which to use. For extremely serious illness, I may use more than one technique. For example, I may begin with conversation and/or healing touch to develop rapport and connect compassionately with the being to be healed; then journey to totems and guides for healing insight and help; and then send a wave of healing energy through reiki. For seriously ill beings, it is my preference to work intensively and sequentially in this way, devoting several periods of time over a couple of days to do one complete healing. After that, my work is done and I simply leave the intention on my altar for a few months of maintenance healing manifestation.

Chapter Ten

Methods of Energy Input, Clearing & Extraction

The following sections discuss four methods of healing. These methods can be done in any order that feels intuitively appropriate for you and your client. Methods of clearing and extraction of energy involve more physical movement with hands than putting positive energy into the clients. Unless the animal or person is very comfortable with you, it may be best to initially focus on the first method of putting positive energy in.

- Methods for Positive Energy Input

- Methods for Energy Field Clearing

- Methods for Negative Energy Extraction

- Methods for Remote Distance Healing

Methods for Positive Energy Input

Putting positive healing energy into the client often focuses on one chakra point at a time. In Reiki of the

Human, the healing practitioner starts from the crown chakra and moves down to the root chakra, and then addresses the limbs. (Rand, Stein) In Healing Touch for Animals, the healing practitioner starts at the root chakra and moves to the crown chakra, and then addresses the limbs. (Komitor) The hands are placed side by side, in a cupped manner, with the center of the palms above the chakra. Reiki Masters usually suggest about 3 minutes of energy healing per chakra and per position along the limbs.

After healing each chakra until the hand feels the position is done, balancing of the chakras can be done by placing one hand on one chakra and another on the next chakra, feeling the flow of energy between the chakra points. With animals, this means placing one hand on the root and the other on the sacral and holding that position for about 3 minutes, then moving upward with the sacral and solar plexus, and so on.

After dealing with the chakra points and limbs, the healer intuitively addresses any location that the hands are taken to for special healing of a diseased organ or joint. If a specific organ is known to be diseased, it may be useful to also conduct healing all along the meridian related to that organ.

Thereafter, the healing may be completed by standing a couple of feet away and holding the hands facing toward the client to sense and heal the entire energy field of the patient's body, willing it to be balanced, flowing freely, clear, energized and protected.

Hand placement varies depending on your intention and according to your intuitive choices at the time. A complete healing often involves use of several hand placements in the sequence listed below:

- **Chakra-by-Chakra.** One healing method involves placing both hands on one chakra and moving through the sequence of chakras, from the root to the crown. This healing method focuses healing energy on one chakra at a time. The healer intuitively stays at one point until feeling ready to move on to another and may decide to return to a specific point at the end of the cycle.

- **Bridging Chakras.** One healing method involves placing one hand on one chakra and the other hand on an adjacent chakra, then step-by-step moving the two hands through two chakras upward from the root to the crown. This healing method focuses on balancing the energy flow between chakras, harmonizing the healing among all the points.

- **Whole Body Whirlpool.** One healing method involves placing one hand on the root and another on the heart and visualizing the energy circulating between the hands and flowing to heal all the chakras. This improves overall well being and is relaxing for an animal that is nervous about you moving your hands from place to place. For me, this method is most relaxing and allows me to focus my intent on a full healing without moving. It is particularly pleasant when working with one's own pets while lying on the couch together.

- **Heart Healing**. One method of healing focuses entirely on the heart chakra and sends energy–healing to the entire body by pushing it through the heart. The two hands are placed side-by-side on the chest of the animal above the heart chakra. Feel the heart energy flowing over into the entire body until every organ and limb has been flooded and flushed with heart energy.

Methods for Energy Field Clearing

- **Raking Energy Field.** To powerfully clear an energy field of negative energy, the healing intuitive rakes spread fingers through the energy field, about 4 to 6 inches from the animal's body. The fingers act like magnets, picking up negative energy. The energy-healing practitioner feels electrical prickling in the fingertips where there are problems. If the energy feels very prickly, raking is done until the energy field is finally cleared.

- In raking to clear energy, the spread fingers grab onto bad energy as they move through the patient's energy field. At the end of the limb or body part being raked, the healer continues the raking about 12 inches from the body's end and then throws the raked negative energy away. The process continues… rake, then discard, rake, then discard.

- For this process to proceed with greater effect there is value to using one's breath to increase the focus. I breathe in as I rake through the energy field and pull out the negative energy, and I breathe out strongly as I discard that energy, shaking my hands strongly to dislodge the energy from my hands.

- If some accumulated prickling or negative energy bothers the healer, the hands are usually shaken until everything is released. Another method of throwing away the negative energy simply involves facing the palms outward and focusing on clean energy pouring out through the hands to the cleansing universe. The healer visualizes being an open channel, like a flexible, open bamboo, through which the clean, universal healing energy flows, unimpeded.

- It is valuable to thank Mother Earth for receiving, assimilating, and healing this negative energy.

- Hands usually need to be washed with cool clean water to stop the prickling sensation. This should be done before proceeding with energy healing.

- **Unruffling the Energy Field**. Where energy appears to be knotted or tangled, it can be loosened and expanded by the unruffling technique. (Komitor) The technique involves using a hand over hand motion, brushing down and away, and brushing back and away, in a continuous flowing movement. The fingers are cupped like claws and separately, much as a large combs teeth would be separated to work through a hair knot. Work your way through the knots with your intention and your fingers mirroring your intention. Follow your intuition. There is no negative energy to discard in this activity. It is a simple untangling.

- **Feather Soothing the Energy Field.** A gentle soothing clearing can be done with a feather or group of feathers, simply using a dusting motion around the fringes of the client's energy, to open up the energy edges toward expansion.

- Sage may be burned and the feather may be used to brush the sage smoke through the energy field around the body and clear it. This is a light touch in clearing, but be sensitive that some clients may find sage smoke unpleasant and prefer that it not be used.

Methods of Negative Energy Extraction.

The above Methods put energy into your client and clear their outer energy field, sometimes referred to as the aura. On some occasions, the healer may wish to do an extraction of disease, rather than inputting healing energy. This process would follow the above steps of intention; however the step involving management of the breath would be different. For extraction, the healer would not want to envision the arms as hollow tubes, as there would be no desire to allow disease into the healer's body. Instead, the breathing would be conducting only in and out of the healer's torso. The healer's arms and hands would be used as tools to grab and extract negative energy and throw it away.

During chakra healing and balancing, an energy-healing practitioner may feel a particular resistance in an area and intuit that it is necessary to untangle energy knots or pull out energy cords. To do that, the healing practitioner may use his or her breath and reach toward the area, pulling the negative energy object outward, throwing it toward the receptive universe. This activity is done repeatedly until the healing practitioner feels it has been effective.

Master training in Usui Reiki teaches a method of removing negative energy objects, using the Violet Breath. The Violet Breath is done by placing the tip of the tongue against the upper palate or roof of the mouth and holding it there with the mouth slightly open. Contracting first the anus and then the muscles of the genitals contracts the Hui Yin area, between the genitals and anus. This contraction closes the circuit for the Ki, so that it flows only in the trunk area. The contraction is held during the entire time that the breath is circulated downward from the Hara, or navel area, upward along the back and down to the navel area again, around and around in a circle during the extraction healing exercise. This requires significant practice and control. Reiki Masters use this same Violet Breath for passing attunements. (Stein)

To remove negative energy, the advanced energy-healing practitioner stands before the client and focuses intent. The practitioner explains the next step of activity, as the breathing will be emphatic. The healing practitioner begins the Violet Breath gently, then with a strong inward Violet Breath reaches toward the client to take out a negative energy object then, with a strong outward Violet Breath, throws that object to the receiving universe. This inward and outward movement is repeated until the negative energy block feels removed or at least lessened. Several extraction healings may be needed for serious disease.

Distance-Healing

Any form of energy healing can be done from a distance. No tools are required except your intent. The simplest thing to do is to ask your higher power, which is your inner source connection to the Great Creator, to do a healing. Never overlook the simple power of prayer.

To focus your intent, you might use a common prayer technique, like saying the rosary, or walking on your knees to a shrine, or marching miles in a pilgrimage to a holy place, or bowing to Mecca. It you wish to focus your intention on a specific body part, you may use a stuffed animal and place you hands on the correct body part, or use a photograph of an animal and put your finger on that body part. Others write an affirmation about the part of the body needing healing. Others make a drawing. Simply envision the client and the part of the body and do the energy healing work exactly as you would if he or she were right there in the room with you.

An Affirmation

My first experience with distance-healing was in Thailand 20 years ago, following more than a year of daily study of *A Course in Miracles*. It was 2AM and I was visualizing healing energy being sent to an elderly man with cancer who was living about 2 miles away. The next morning I saw him and he said, "What did you do to me last night at 2AM? I felt all the pain in my body go away and I sensed that you were there. This morning, for the first time, I didn't need any pain medication."

I was stunned. This was my first major affirmation that distance-healing was not just in my mind. I had said nothing about my interests in healing to this man and felt blessed that he experienced a healing strongly enough to comment. He was a highly successful businessman and with multiple large retail establishments around the world and I was on an engineering assignment for a multi-lateral donor agency, and we had had no conversation about spiritual or energy aspects of life. It was a wonderful affirmation for us both, and he gave me a relic that I shall always treasure.

During the study of Reiki, one of the Usui Reiki Level II attunements is a distance-healing attunement. It is provided to open the channel to enable distance healing. Distance-healing is comparable to the age-old traditions of sending love and sending prayers. Through training, sending energy healing is strongly experienced because of the healing practitioner's ability to focus intent.

In distance-healing of animals, I have found it useful to talk with the owners on the telephone and have them put their hands on the body parts as I hold my hand out in front of me and send energy to the person to transmit to the animal. I have also found it useful for an animal owner to touch the animal in a given place, while I am touching that part of my body, and we are in touch by phone. Some people use stuffed animals to focus their intention to specific body parts during distance healing. Thus the healing power is heightened by the empowerment of others during such a two-person partnership of distance-healing.

Ending an Energy-Healing Session.

If the healing practitioner suddenly removes his or her hands and abruptly ends a session, it feels like a shock to the client. Always move from one healing position to another with slowness, preferably moving one hand to the new position and then bringing along the other hand.

As you close the session, visualize the energy field of the client as whole and collected, and use your hands to gather it toward the client and seal it up, as if zipping it up.

At the end, from behind the client, put first one hand on the shoulders, and then the other. Stand quietly and make an affirmation for the healing session to be successfully concluded, for the client to have all of their energy chakras protected, and for the healing to continue on its own to bring lasting effects. Then offer the client a drink of cool fresh water, and encourage them to remain quiet and still for a few minutes.

Descriptions of Earth Lodge®
Energy Healing for Animals
Level I, II, III and Master Training Courses

Level I — one and a half days

Receipt of Level I Usui Reiki attunement (Cho-Ku-Rei symbol to open the healing channel and increase power), Level I training on energy healing for animals, exercises to sense energy fields and pick up intuitive information, exercises to diagnose energy blocks, meditations to gather clear and gather energy, exercises to focus intent and use the breath in healing work and energy-healing practice on dogs.

Level II — one and a half days

Receipt of Level II Usui Reiki attunement (Sei-He-Ki symbol to purify and stimulate emotional healing, and Hon-Sha-Ze-Sho-Nen symbol for distance and timeless healing), Level II training on energy healing for animals, exercises in remote sensing, meditations to clear and gather energy, distance healing, extraction and continued basic training in co-creative energy healing using your intent and breath, and energy-healing practice on horses.

Level III and Master — two days

Receipt of Level III and Master Usui Reiki attunements (Tibetan Fire Serpent symbol, also called the Raku, which cleanses and magnifies healing, much like the breath of the Mayan God Quezecolte, and the Tibetan Dai-Ko-Mio Master symbol and the Usui Master Dai-Ko-Myo symbol which both strengthen all previous levels of attunement), learning to write the Level I, II, III and Master Reiki symbols over your hands to strengthen your healing, practice in the Violet Breath, energy block extraction training, journeying for healings, completion of all requirements for Earth Lodge Energy Healing for Animals Level III and Master certification.

Note: This series of courses uses a variety of energy-healing techniques and includes the Reiki attunements. However, to fully understand Reiki and become a certified Reiki practitioner for humans, separate Reiki training is required and involves work on humans at each level.

Useful References

1. Adam, Michael, *Wandering in Eden – Three Ways to the East within Us*, Alfred A. Knopfe Publishers, New York, USA, 1976.

2. Angel, Jeremy, *Cats Kingdom: A Long and Loving Look into a Very Special Feline Community*, Warner Books, New York, USA, 1987.

3. Ashcroft-Nowicki, Dolores and Brennan, J.H., *Magical Use of Thought Forms: A Proven System of Mental and Spiritual Empowerment*, Llewellyn Publications, Minnesota, 2006

4. Bear Heart with Molly Larkin, *The Wind is My Mother: The Life and Teachings of a Native American Shaman*, A Berkley Book with Crown Publishers, Inc., New York, USA, 1996.

5. Brennan, Barbara Ann, *Hands of Light: A Guide to Healing Through the Human Energy Field*, Bantam Books, New York, USA, 1987.

6. Budiansky, Stephan, *The Nature of Horses: Exploring Equine Evolution, Intelligence and Behavior*, The Free Press, New York, USA, 1997.

7. Budiansky, Stephan, *Truth about Dogs, The : An Inquiry into Ancestry Social Conventions Mental Habits Moral Fiber Canis fami*, Penguin Books, New York, USA, 2000/2001.

8. Busch, Robert H., *The Wolf Almanac: A Celebration of Wolves and Their World*, Connecticut, USA, 1995/1998.

9. Calhoun, Peter, *Soul on Fire: A Transformational Journey from Priest to Shaman*, Hay House Publishers, California, USA, 2006.

10. Castaneda, Carlos, *Magical Passes: The Practical Wisdom of the Shamans of Ancient Mexico*, Harper Collins, New York, USA, 1998.

11. Cointreau, Maya and E. Barrie Kavasch, *Equine Herbs & Healing: An Earth Lodge Guide to Horse Wellness*, Earth Lodge Publications, 2007.

12. Cointreau, Maya, *Grounding and Clearing: Being Present in the New Age*, Earth Lodge Publications, 2008.

13. Cointreau, Maya, *Natural Animal Healing: An Earth Lodge Guide to Pet Wellness*, Earth Lodge Publications, 2009.

14. Cointreau, Maya, *To the Temples: 14 Meditations for Healing and Guidance*, Earth Lodge Publications, 2007.

15. Coren, Stanley, *How Dogs Think: Understanding the Canine Mind*, The Free Press, New York, USA, 2004

16. Coren, Stanley, *How to Speak Dog: Mastering the Art of Dog-Human Communication*, The Free Press, New York, USA, 2000.

17. Dalai Lama, *The Little Book of Wisdom*, Rider, London, 2000.

18. Donaldson, Jean, *The Culture Clash: A Revolutionary New Way of Understanding the Relationship between Humans and Domestic Dogs*, James & Kenneth, California, USA, 1996.

19. Feng, Gia-Fu and Jane English, *Tao Te Ching*, Vintage Books, Random House, New York, USA, 1972.

20. Fisher, John, *Think Dog: An Owner's Guide to Canine Psychology*, Trafalgar Square Publishing, Vermont, USA, 1995/2003.

21. Foundation for Inner Peace, *A Course of Miracles*, California, USA, 1975.

22. Fulton, Elizabeth and Prasad, Kathleen, *Animal Reiki*, Ulysses Press, California, 2006.

23. Gawani Pony Boy and Essays of Various Horsewomen, Of Women and Horses, Bow Tie Press, California, USA, 2000.

24. Gawani Pony Boy, *Horse, Follow Closely: Native American Horsemanship*, Bow Tie Press, California, USA, 1998/2003.

25. Grimassi, Raven, *The Witches Craft – The Roots of Witchraft and Magical Transformation, Llewellyn Publications*, Woodbury, Minnesota, 2002.

26. Hale, Gill; Stella Martin and Josephine de Winter, *The Feng Shui Home: Creating Spiritual Spaces in your Environment with Color, Altars, and Shrines, Space Clearing and the Ancient Chinese Art of Feng Shui*, Hermes House, London, UK, 2004.

27. Harner, Michael, PhD/Anthropology, *The Way of the Shaman*, Harper and Row, California, USA, 1980/1990. (http://www.shamanism.org)

28. Hempfling, Klaus Ferdinand, *Dancing with Horses: The Art of Body Language* (translated from German), Trafalgar Square Publishing, Vermont, 2001/2003.

29. Hicks, Ester and Jerry Hicks, *The Law of Attraction*, Hay House, California, USA, [].

30. Hourdebaigt, Jean Pierre, *Equine Massage: A Practical Guide*, Howell Book House, New York, USA, 1997.

31. Ingerman, Sandra, *Soul Retrieval: Mending the Fragmented Self*, Harper Collins Publishers, California, USA, 1991.

32. Kavasch, E. Barrie, *The Medicine Wheel Garden: Creating Sacred Space for Healing, Celebration and Tranquility*, Bantam Books, New York, USA, 2002.

33. Komitor, Carol, *Healing Touch for Animals: Level 1*, Komitor Healing Method, Inc., Colorado, USA, 2000. (http://www.healingtouchforanimals.com)

34. Lao Tzu (translated by D.C.Lau), *Tao Te Ching*, Penguin Books, Middlesex, England, 1963.

35. Mails, Thomas E., *Fools Crow: Wisdom and Power*, Council Oak Books, Arizona, USA, 1991.

36. Mails, Thomas E., *Secret Native American Pathways: A Guide to Inner Peace*, Council Oak Books, Arizona, USA, 1991.

37. McTaggert, Lynne, *The Field: The Quest for the Secret Force of the Universe*, Harper, New York, (revised edition, 2008)

38. Mech, L. David, et.al., *The Wolves of Denali*, University of Minnesota Press, Minnesota, USA, 1998.

39. Mech, L. David, *The Wolf: The Ecology and Behavior of an Endangered Species*, University of Minnesota Press, Minnesota, USA, 1970/1981.

40. Medicine Grizzlybear Lake, Native Healer: Initiation into an Ancient Art, Quest Books, Illinois, USA, 1991.

41. Moss, Robert, *Conscious Dreaming: A Spiritual Path for Everyday Life*, Random House, New York, USA, 1996.

42. Motz, Julie, *Hands of Life*, Bantam Books, New York, USA, 1998.

43. Putoinen, C. J. , *Natural Remedies for Dogs and Cats*, Keats Publishing, USA, 1999.

44. Rand, William Lee, *Reiki Master Manual, Including Advanced Reiki Training*, International Center for Reiki Training, Michigan, USA, 2003. (http://www.reiki.org)

45. Rand, William Lee, *Reiki The Healing Touch: First and Second Degree Manual*, Vision Publications, Michigan, USA, 1991/1998.

46. Rashid, Mark, *Horses Never Lie: The Heart of Passive Leadership*, Johnson Printing, Colorado, USA, 2000.

47. Roberts, Monty, *The Man Who Listens to Horses*, A Ballantine Book, USA, 1996/1997.

48. Rugaas, Turid, *On Talking Terms with Dogs: Calming Signals*, Hanalei Training Center, Inc., Washington, USA, 1997.

49. Silvester, Hans, *Horses of the Carmargue*, Harry N. Abrams, Inc., New York, USA (translated from original French book), 2003.

50. Simpson, Liz, The Book of Chakra Healing, Sterling Publishing, New York, 1999.

51. Snow, Amy and Nancy Zidonis, *The Well-Connected Dog: A Guide to Canine Acupressure*, Tallgrass Publishers, Colorado, USA, 1999.

52. Stein, Diane, *Essential Reiki: A Complete Guide to an Ancient Healing Art*, The Crossing Press, California, USA, 1995/1999.

53. Watts, Alan W., *The Way of Zen*, Vintage Books, Random House, New York, USA, 1957.

54. Wesselman, Hank, PhD/Anthropology, *The Journey to the Sacred Garden: A Guide to Traveling in the Spiritual Realms* (includes CD of drumming), Hay House, Inc., California, USA, 2003. (http://www.sharedwisdom.com)

55. Wesselman, Hank, PhD and Jill Kuykendall, RPT, Spirit Medicine: Healing in the Sacred Realms, Hay House, Inc., California, USA, 2004.

56. White Eagle, *The Quiet Mind*, The White Eagle Publishing Trust, Liss, England, 1972.

57. Williams, Marta, *Learning Their Language: Intuitive Commication with Animals and Nature*, New World Library, California, 2003.

58. Wright, Machaelle Small, *Behaving as if the God in All Life Mattered*, Updated and Revised Version, Perelandra, Virginia, 1997.

59. Zidonis, Nancy; Amy Snow and Marie Soderberg, *Equine Acupressure: A Working Manual*, Tallgrass Publishers, Colorado, USA, 1999

Dog Meridians & Acupressure Points

Heart Meridian

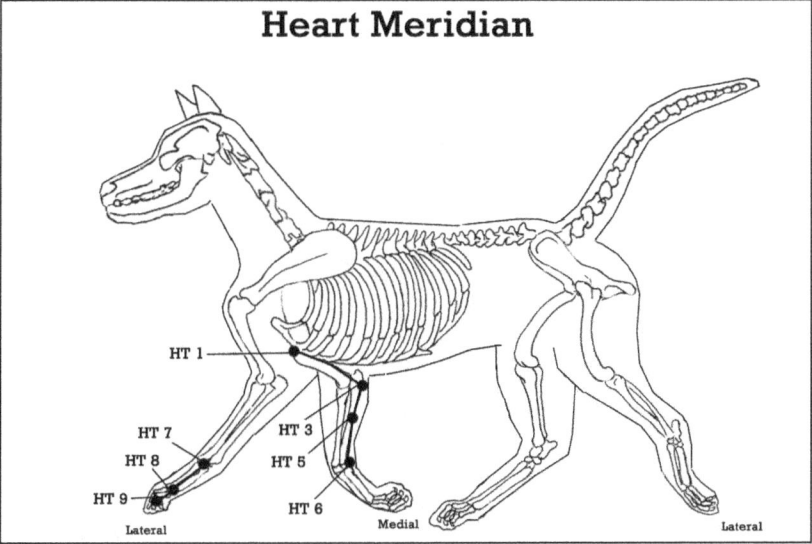

HT 1
HT 7
HT 8
HT 9
HT 3
HT 5
HT 6
Lateral
Medial
Lateral

Liver Meridian

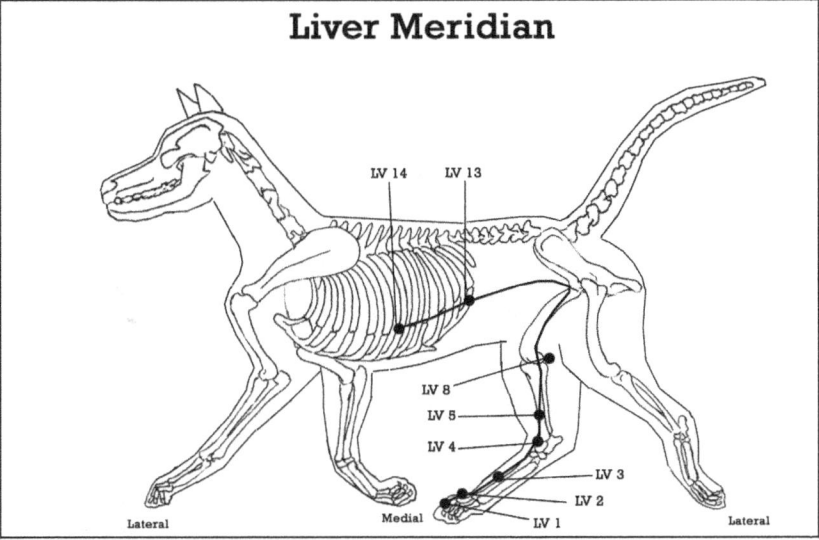

LV 14
LV 13
LV 8
LV 5
LV 4
LV 3
LV 2
LV 1
Lateral
Medial
Lateral

Lung Meridian

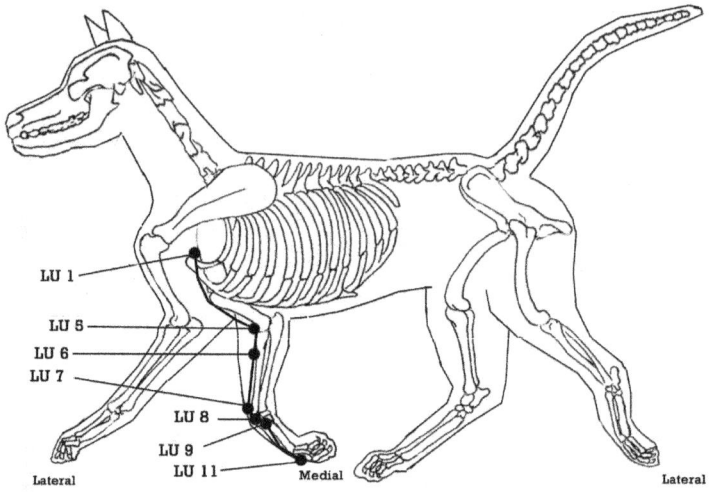

LU 1
LU 5
LU 6
LU 7
LU 8
LU 9
LU 11

Lateral   Medial   Lateral

Kidney Meridian

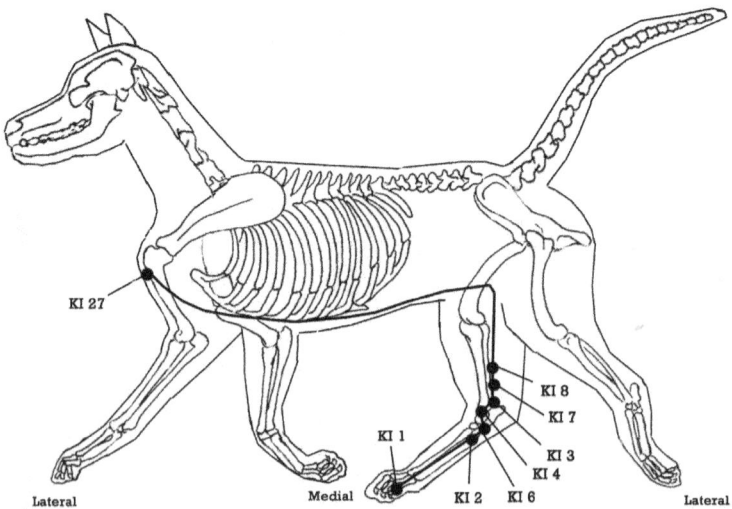

KI 27

KI 8
KI 7
KI 1
KI 3
KI 4
KI 2   KI 6

Lateral   Medial   Lateral

Gall Bladder Meridian

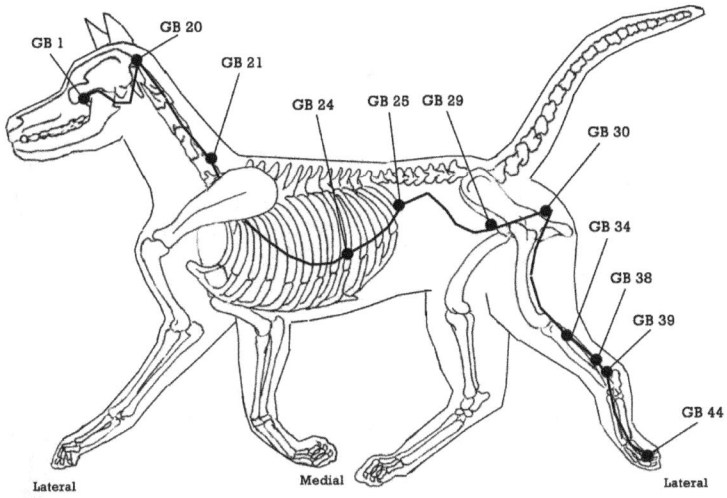

GB 1
GB 20
GB 21
GB 24
GB 25
GB 29
GB 30
GB 34
GB 38
GB 39
GB 44

Lateral
Medial
Lateral

Stomach Meridian

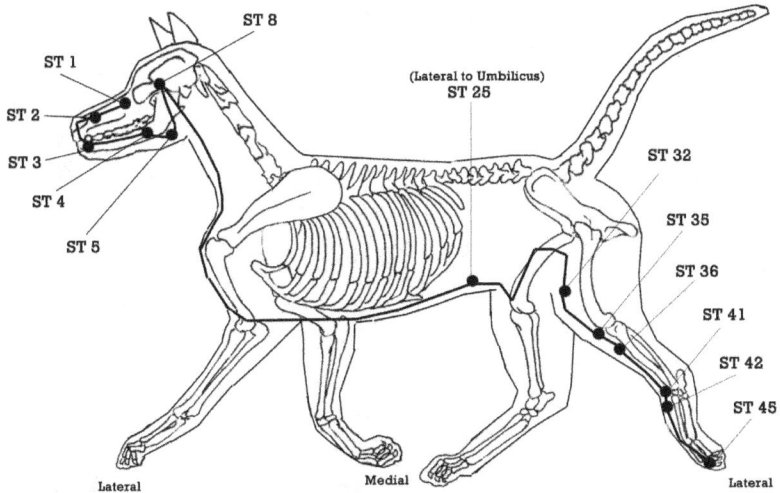

ST 8
ST 1
ST 2
ST 3
ST 4
ST 5
(Lateral to Umbilicus)
ST 25
ST 32
ST 35
ST 36
ST 41
ST 42
ST 45

Lateral
Medial
Lateral

Pericardium Meridian

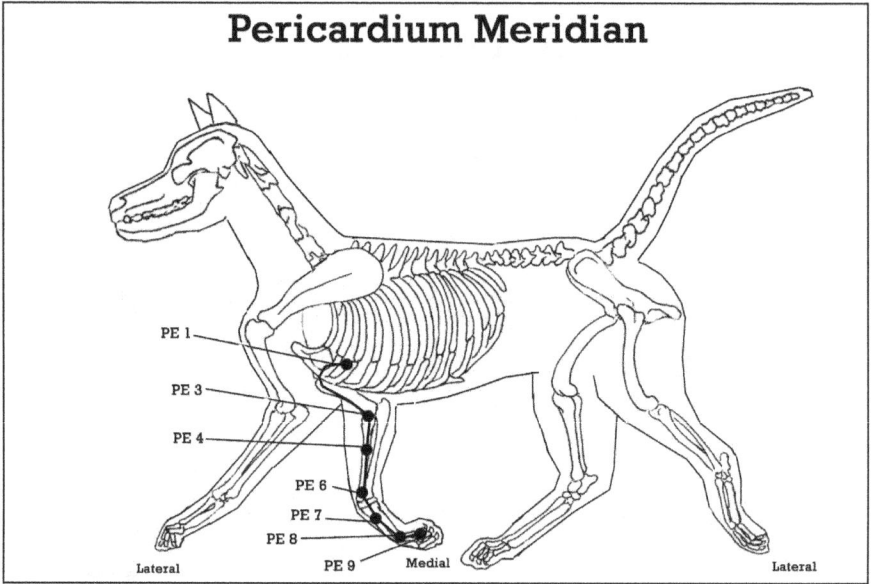

PE 1
PE 3
PE 4
PE 6
PE 7
PE 8
PE 9

Lateral
Medial
Lateral

Spleen Meridian

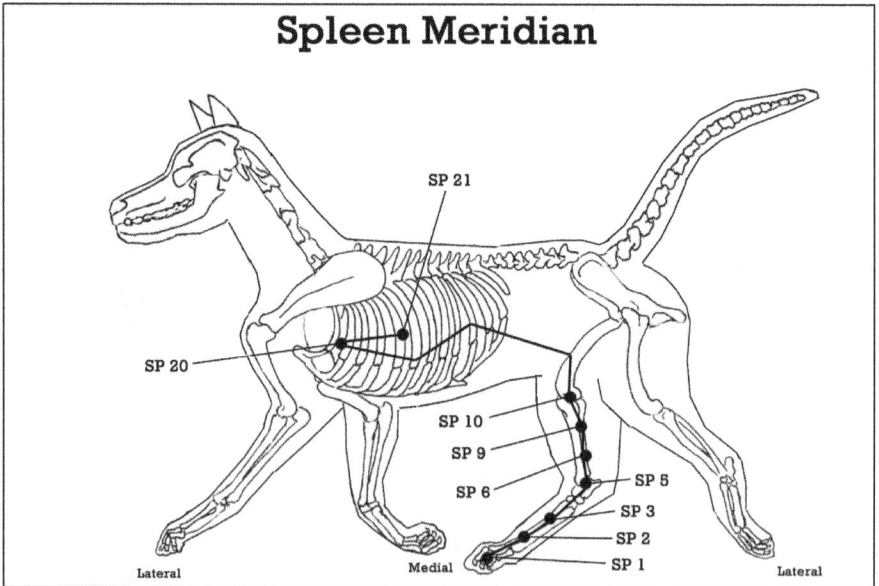

SP 21
SP 20
SP 10
SP 9
SP 6
SP 5
SP 3
SP 2
SP 1

Lateral
Medial
Lateral

Large Intestine Meridian

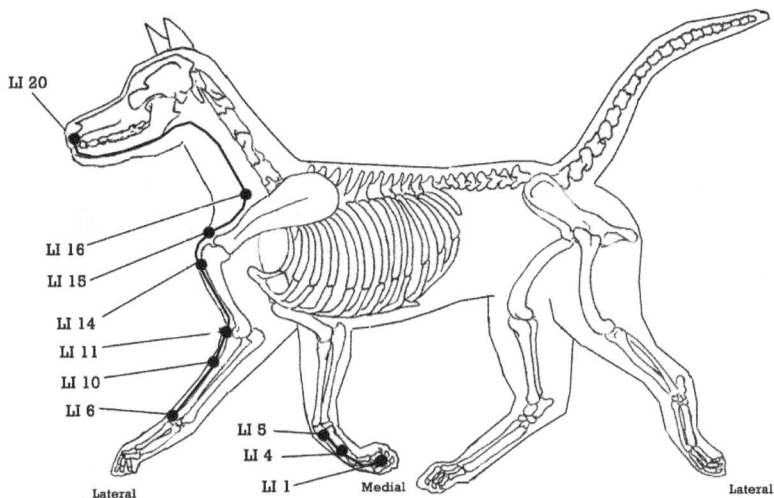

LI 20

LI 16
LI 15

LI 14
LI 11
LI 10
LI 6

LI 5
LI 4
LI 1

Lateral

Medial

Lateral

Small Intestine Meridian

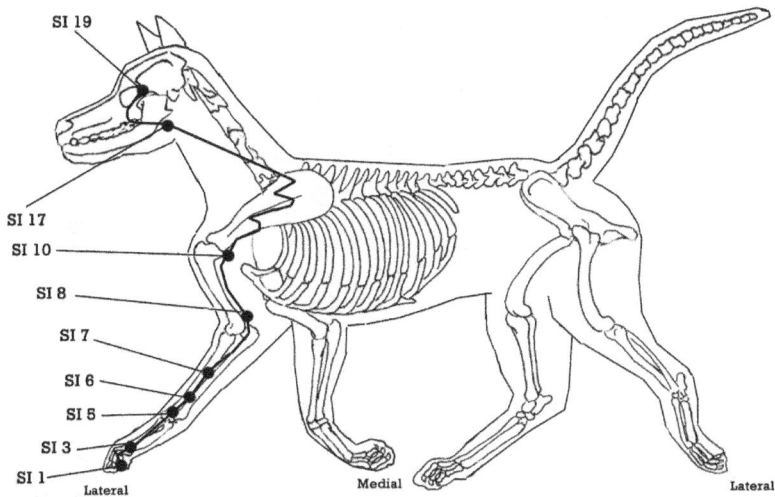

SI 19

SI 17
SI 10

SI 8

SI 7

SI 6
SI 5

SI 3
SI 1

Lateral

Medial

Lateral

Conception Vessel Meridian

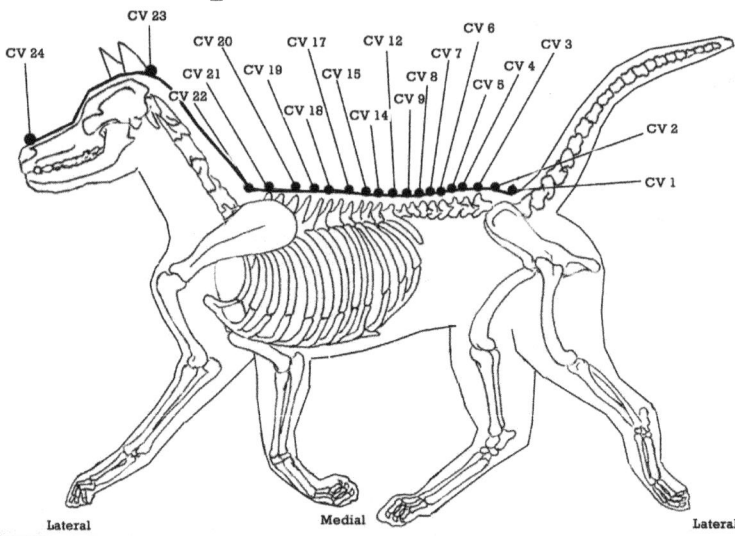

CV 24 · CV 23 · CV 22 · CV 21 · CV 20 · CV 19 · CV 18 · CV 17 · CV 15 · CV 14 · CV 12 · CV 9 · CV 8 · CV 7 · CV 6 · CV 5 · CV 4 · CV 3 · CV 2 · CV 1

Lateral · Medial · Lateral

Bladder Meridian

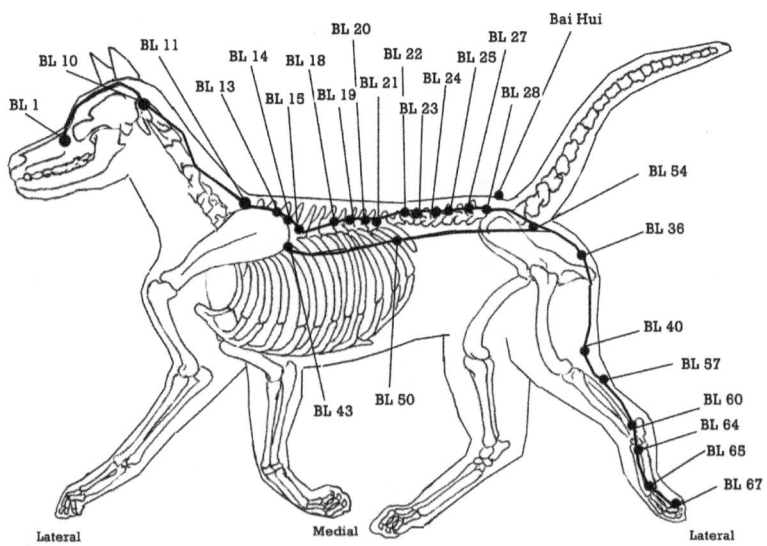

BL 1 · BL 10 · BL 11 · BL 13 · BL 14 · BL 15 · BL 18 · BL 19 · BL 20 · BL 21 · BL 22 · BL 23 · BL 24 · BL 25 · BL 27 · BL 28 · Bai Hui · BL 54 · BL 36 · BL 40 · BL 57 · BL 43 · BL 50 · BL 60 · BL 64 · BL 65 · BL 67

Lateral · Medial · Lateral

About the Author and Teacher

For 40 years, Sandra Cointreau has worked as an engineer on civil/environmental engineering and planning projects. She specializes in waste management systems and has worked in the US and over 50 developing countries on municipal, hazardous, medical and livestock waste systems, from collection to disposal and recycling.

Her work on livestock production and processing wastes has their relation to the chain of disease development for livestock-related Zoonotic diseases such as SARS, Highly Pathogenic Avian Influenza, and Transmissible Encephalapathies such as Mad Cow Disease, and their relationship to the growing global resistence to antibiotics related to livestock daily feeding of antibiotics for growth promotion. Her work has also included focus on how to improve the municipal live market and slaughtering facilities that prevail in serving the meat supply for the majority of urban dwellers in developing countries and the humane treatment of animals at these facilities.

Sandra's parallel work with animals and energy healing has provided intuitive and creative balance to harmonize the analytical demands of engineering. She has bred over 100 standard poodle puppies and cared for her own horses, cats, rabbits and fish at her country home, Stoney Brook Farm, in Roxbury, CT. She also paints watercolors, often of animals, and particularly of horses.

For training in energy healing, she has completed the following:

- A Course in Miracles, four years of daily intensive study and meditation, 1986-1990.

- Shamanism, drumming, and space clearing studies with Native American traditional healing and ethnobotany expert, E. Barrie Kavasch, Medicine Wheel Wellness Center, 1998-2002.

- Shamanism, Level I, Michael Harner, Foundation for Shamanic Studies, 1999.

- Komitor Healing Touch® for Animals, Level I, Carol Komitor, 2003.

- Shamanism, Healing the Shadow Self and Solstice Vision Quest, Paul Sivert, 2007.

- Usui Reiki I and II, Catherine Burke, Usui Reiki Master and Certified Massage Therapist, 2003.

- Usui Reiki III and Reiki Master/Teacher, William Lee Rand, International Center for Reiki Training, 2004.

- Karuna Reiki® I, II, III and Master/Teacher, William Lee Rand, International Center for Reiki Training, 2008. (Karuna Reiki® is taught only to Usui Masters, adding a total of 9 new Karuna Reiki® symbols for deeper and more focused healing potential).

About Reiki

Reiki is traced to Japan and Tibet, and is noted in ancient documents more than 2000 years old. In the late 1800's, Dr. Mikao Usui, a theologian from Kyoto, Japan, rediscovered this healing method. He studied Indian Sanskrit and found the Sanskrit Buddhist symbols and information in Japanese monastery archives. These drove him to meditate alone on the holy mountain of Kuriyama, where he obtained enlightenment on each symbol and how to use it.

It is the tradition for all certified Usui Reiki healing practitioners to record and state their lineage from Dr. Usui. When he died in 1930, he left 16 Reiki Grand Masters able to continue his work. Most western Reiki healing practitioners have their lineage through a Japanese-American woman, Mrs. Hawaya Takata, from Hawaii, who traveled to Japan for training by one of Usui's Grand Masters.

William Lee Rand, founder and Director of the International Center for Reiki Training in Detroit, Michigan has been trained in Japan through traditional Usui Reiki masters, as well as in the USA. He is also the contemplative creator, together with a team of leading

Reiki Masters, of the highly powerful additional levels of Karuna Reiki ®, which add an additional 9 Reiki symbols for different uses and an overall significantly more intensive healing ability.

Note: Karuna is a Sanskrit word that means compassionate action coming from an unbounded sea of love.

Blank Pages for Journaling — Your Notes

Blank Pages for Journaling — Your Notes

NEW IN 2009

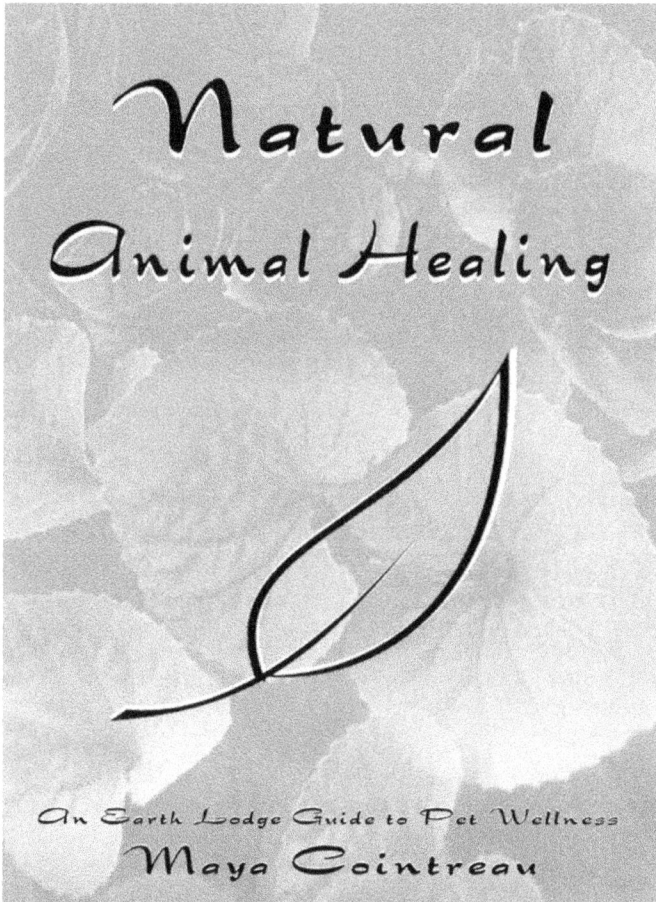

Natural Animal Healing
An Earth Lodge Guide to Pet Wellness
by Maya Cointreau
Paperback, 164pp, $19.95

This exciting new guide from Earth Lodge includes natural health solutions for pets from many modalities including over homeopathy, flower essences, energy healing, animal communications, aromatherapy, crystal healing and over 50 pages of herbs and a comprehensive table of ailments and corresponding remedies.

www.earthlodgeherbals.com

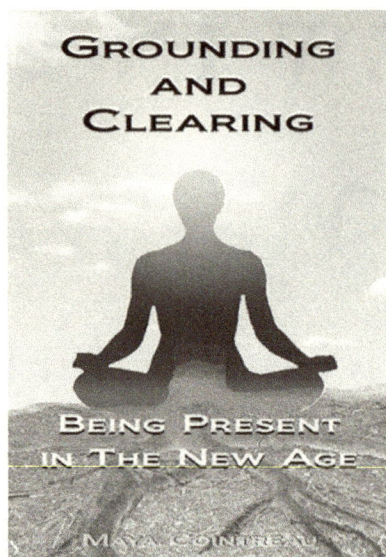

Grounding & Clearing: Being Present in the New Age
by Maya Cointreau, Copyright © 2008
Paperback, 136 pages, $18.50

Grounding & Clearing: Being Present in the New Age gives you the tools you need to remain a focused and empowered channel for your higher self, allowing you to manifest the reality you desire. In the New Age, we spend much of our time and energy looking for new ways to open the third eye, to heighten our awareness, to ascend to new levels of healing. Many of us use affirmations and the law of attraction to create a more positive future. However, we can only birth our new reality if we also remain grounded in the physical. In order to receive, embody and enact the messages that our higher selves send us, our bodies must be strong and aligned right along with our chakras and our souls.

There is no mystery to the grounding techniques detailed in Grounding & Clearing. In this book you will learn techniques to ground in any situation, and to clear negative patterns and energies from your life. You will learn how to ground with prayer, scents, candles, symbols, colors, breath, nature, and more. With regular grounding and clearing, you will remain calm and focused while you free your spiritual gifts.

"This book can be used by the advanced spiritual practitioner and also by someone with no knowledge of grounding or energy. Each chapter has a different method of grounding/clearing so you are bound to find something new to try! I love this book and will keep it as a reference forever. I highly reccomend this to anyone who ever feels spaced out!"
Rachel Andrews, Professional Medium/Psychic Teacher and Facilitator.

TO THE TEMPLES

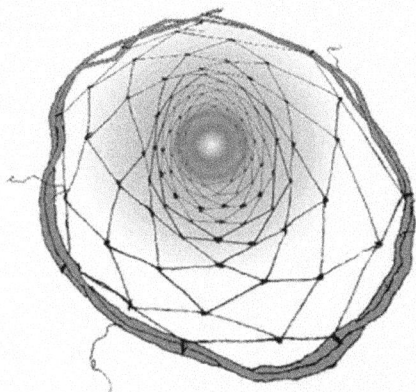

14 Meditations for Healing & Guidance
by Maya Cointreau

To The Temples
14 Meditations for Healing & Guidance
by Maya Cointreau, Copyright © 2007
Paperback, 130pages, $18.50

The fourteen guided meditations in this book are designed to take you on a journey to temples and holy places that exist both in and out of this reality, places without time constraints, preconceived ideas or limitations. Each journey takes you through healing: meeting your guides, native american goddesses, and new teachers; clearing your chakras; and visiting past lives.

These meditations were designed by Maya Cointreau, a shamanic energy healer and herbalist, to help those on the path of healing, whether it be to heal oneself or to heal others, whether you are new to meditation or not. Following each meditation in the book you will also find four beautifully-lined journal pages to record your thoughts and visions.

Available today at www.earthlodgeherbals.com

An Earth Lodge Guide to Horse Wellness

Equine Herbs & Healing

Maya Cointreau & E. Barrie Kavasch
with Sandra Cointreau

Foreword by Allen M. Schoen, MS. DVM

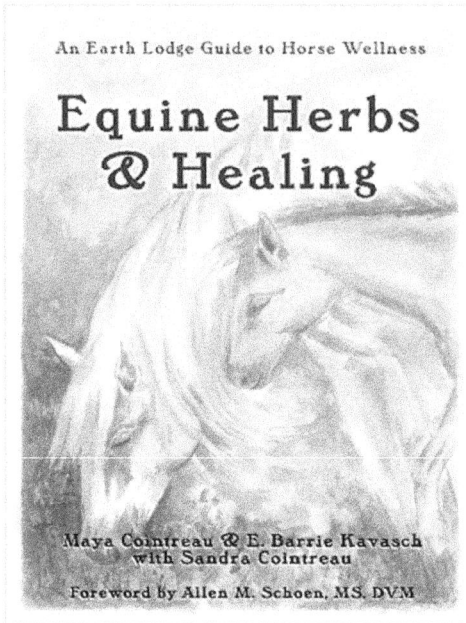

"Equine Herbs & Healing is a must-have resource."
Equine Wellness Magazine

"A great gift."
Natural Horse Magazine

Equine Herbs & Healing
An Earth Lodge Guide to Horse Wellness
by Maya Cointreau & E. Barrie Kavasch with Sandra Cointreau
(Copyright © 2005)
Foreword by Allen M. Schoen, MS, DVM
$19.95

Horses of the past were free to roam on large acreages and commonly sought out the wild herbs and other native medicinal plants they needed to stay properly conditioned. Modern horses rely on their human owners to supply the herbs they need to keep their bodies strong and healthy. The herbalists at Earth Lodge Herbals (www.earthlodgeherbals.com) have brought together years of herbal experience to bring you this Earth Lodge Guide to Horse Wellness: Equine Herbs & Healing, giving you all the tools you need to maintain your horse the natural way.

This informative book teaches you how to combine and use herbs most effectively for your horse's benefit. Learn what herbs have been used traditionally for which ailments and how to make your own salves, tinctures, braces, and sprays. The authors have included a handy reference table of disorders and their corresponding herbal remedies, as well as online resources and herbal recipes for the barn and home. Equine Herbs & Healing covers horse herbalism in all its forms, from the historical uses of dried herbs to advances in aromatherapy and herbal cancer therapy.

www.ingramcontent.com/pod-product-compliance
Lightning Source LLC
Chambersburg PA
CBHW021333090426
42742CB00008B/586